"Hello, Central?"

"Hello, Central?"

Gender, Technology,
and Culture
in the Formation
of Telephone Systems

MICHÈLE MARTIN

McGill-Queen's University Press
Montreal and Kingston • London • Buffalo

© McGill-Queen's University Press 1991
ISBN 0-7735-0830-9

Legal deposit second quarter 1991
Bibliothèque nationale du Québec

Printed in Canada on acid-free paper

This book has been published with the help of a grant
from the Social Science Federation of Canada, using
funds provided by the Social Sciences and Humanities
Research Council of Canada.

All photographs courtesy of Bell Canada Telephone
Historical Collection.

Canadian Cataloguing in Publication Data

Martin, Michèle, 1944–
    Hello Central?
    Includes bibliographical references and index.
    ISBN 0-7735-0830-9
    1. Telephone–Ontario–History. 2. Telephone–Quebec
(Province)–History. I. Title. II. Title: Technology, culture
and gender in the development of telephone systems,
1878–1920.
    HE8864.M37 1991   384.6'082   C91-090009-4

This book was set in Sabon 10/12 by Caractéra inc.,
Quebec City.

À ma mère Nélida,
qui m'a montré le chemin
de la détermination et de la résistance.

Contents

Illustrations 81–90

Tables and Figures ix

Acknowledgments xi

Introduction 3

1 "Freedom Can't Be Overheard" 14

2 Killing the Competition 27

3 The Making of the Perfect Operator 50

4 Voicing the "Pulse of the City" 91

5 Bridging the Gap between the Victorian
  and Modern Eras 110

6 The Culture of the Telephone 140

Conclusion 167

Notes 175

Bibliography 189

Tables and Figures

TABLES

1 Development of the Telephone in Selected Canadian Cities, 1910   35

2 Free Connections in Montreal, 1914   42

3 Proposed Wages and Pay Schedules, 1907   78

4 Wages for Five-Hour Schedule and Overtime, 1907   79

FIGURES

1 Bell's Long-distance Telephone System in Central Canada, 1898   34

2 Montreal Wards, 1881–91   111

3 Social Distribution of Telephone Exchanges in Montreal, 1903   117

4 Bell Telephone System in Montreal, 1905   118

5 Bell Telephone System in Montreal, 1910   119

6 Bell Telephone System in Montreal, 1915   120

7 Bell Telephone System in Montreal, 1920   121

8 Bell Telephone System in Toronto, 1900   122

9 Bell Telephone System in Toronto, 1924   123

10 Telephones per 100 Persons in Selected Countries, January 1921   156

11 Telephone Conversations per Capita in Selected Countries, December 1920   157

Acknowledgments

I am grateful to Alison Prentice, Philip Corrigan, Jack Wayne, and Graham Knight, who, each in her or his particular way, were intellectually stimulating and very supportive. I would like to thank Bell Canada, particularly Stephanie Sykes, section manager of Bell Canada's Historical Records in Montreal, who put no restrictions on my exploration of that company's archives, and provided me with research assistants, working space, and an unlimited photocopying budget. I also owe a debt to Monique Montbriand, of the Diocese of Montreal Archives. Finally, Bruce Curtis was always ready to share my intellectual curiosity as well as the domestic chores, and to give his affectionate support in periods of both profound discouragement and euphoria.

I am also grateful to the journals *Labour/Le Travail*, the *Journal of Communication Inquiry*, and the *Canadian Journal of Communication*, in which some of the material in this book appeared, in somewhat different form.

"Hello, Central?"

Introduction

The Hamilton *Spectator*, in 1884, called the telephone "the youngest and most wonderful development of the means of communication" ("Mr Fawcett and the Telephone" 1884, 1167). Some years later, Casson praised the telephone as "the high-speed tool of civilisation ... the symbol of natural efficiency and co-operation" (1910a, 238). This enthusiasm did not fade during the 1920s. In 1929, Rhodes wrote that the development of the telephone was "a unique and revolutionary step in man's progress" (1929, ix). Despite such contemporary assessments of the significance of the telephone as a factor in social change, this means of communication has often been dismissed as trivial. No serious sociological study has yet been done of the impact of the telephone on society and on social life. This book is about the development of the telephone and its influence on Canadian society. Particular attention is given to the role of women as producers and consumers of telephonic communications and to the consequences of women's activities in shaping this communication technology.

Existing studies of the development of the telephone are generally characterized by a lack of recognition of the importance of gender differentiation in the development of a technology now found in every household (e.g., Brooks 1974, Ogle 1979, Pool 1977). Women, it seems, have had no part in any element of development of this means of communication, except as ignorant and inconsiderate users. This suggests that women passively accepted the telephone system, a common view of women's relationship to new technologies. Very little thought has been given to the contribution of women to the development of means of communication in general.[1]

It seems that researchers have implicitly accepted Cockburn's (1986) claim that men have a monopoly over the development and use of technology. This implies that women are passive witnessess to technological development and accept the uses of technology that men create for them. Analyses of household and reproductive technologies, for instance, like those of communication technologies, are generally empty of analyses of women's participation in technological development. Although these studies discuss such important issues as the impact of technology on women's lives and on the household (Conseil du statut de la femme 1988, Rothschild 1983, Whicker 1985), they rarely look at the other side of the relationship: the influence of women's practices on the development of technology. Moreover, relatively little thought is usually given to the political-economic development of the technology.

This book takes a different perspective, aiming to show that women make an active contribution to the development of certain technologies. I have drawn upon the rich and largely unexplored archives of Bell Canada in Montreal to examine the development of the telephone system in Canada, especially in Ontario and Quebec, with some reference to the United States, over the period from the invention of the telephone to the system's automation (1876 to 1920). I stress the relationship between political and economic factors in the development of a new technology, the creation of a new occupation for women workers, and the development of new cultural practices.

I examine the historical impact of a new technology on cultural practices, especially on social relations among women. Unlike the few existing descriptive studies of the telephone, this book is concerned with the interaction between private capitalist interests in the telephone industry and the cultural practices of women users in the formation of a specific telephone system. It examines women's participation in the development of the telephone as consumers and as operators, and the influence they had on the expansion of the telephone system and on its structure. It details the possibilities for several alternative kinds of telephone systems, and demonstrates that machines are neither neutral instruments which can be used any which way, nor simply deterministic technologies. Rather, technologies have a valence: a limited range of uses are possible,[2] and conflicts between developers and consumers determine which use or uses will predominate.

Although there are works that examine the effect of technology on women at work or in the household, most are concerned neither with the subjection of women to the machine, nor with the influence of two-way communication technology on women's lives. I examine in detail the steps that led to the subjection of operators to their machines, and the various practices adopted by women in a developing telephone culture.

Women's contributions to the development of forms of communication have generally been oblique, consisting of positive, active practices, not always consciously oppositional, but consciously developed to suit their own activities. These practices take advantage of lacunae in planned development, enabling women to counter not only the prescribed uses of the technology, but its planned pattern of distribution as well. Women have not had the power to create forms of communication; rather, they are forced into a "defensive resistance," to use A. Mattelart's words, since there seems to be no other way to make the changes they want in social practices.

According to Walkerdine (1985), in a patriarchal capitalist society women face a paradoxical situation in which they have only two choices: either they agree to participate in the existing form of power and have a chance to validate themselves, or they reject male domination in order to learn more about their own characteristics and thus are devalued. In the first case, by agreeing to reproduce male power and social control, women lose their gender particularities and become only a part, or shadow, of their sexual counterparts. In the latter case, women can live according to their own characteristics, albeit consenting to lose power and control over social change. Confronted with such a situation, says Walkerdine, women often keep silent as a defence against either option.

In the case of this study, women's reaction to the lack of power and control was the opposite of silence.[3] Indeed, it was women's extensive use of the telephone that eventually forced the industry to change some of its plans. Women were resisting the proposed order; this resistance, though defensive in the sense that someone was attacking the rationality of their social practices, was nonetheless active and efficient: women's activities opposed telephone companies' prescribed uses of the telephone. I should stress here, to preclude giving a romantic connotation to "resistance" and "opposition," that they involved concrete actions and practices performed by women in their daily lives, which contradicted expected uses imposed by the men who controlled the system, and which varied according to the class to which the women belonged and the position they occupied in society. This resistance was not generally translated into spectacular actions purposely directed against the telephone companies. It was mainly unconscious, and came from more or less small, independent groups. But it forced some important changes upon the planned pattern of distribution and uses of telephone systems.

To identify this form of participation by women in society, it is essential to look at the historical conditions within which it occurred. A feminist analysis would reveal women's participation in terms of roles, but might not link their practices with the political-economic conditions within which they were created. On the other hand, a political-economic analysis, which would stress these conditions, might neglect the partic-

ipation of women in the elaboration of systems of communication, insofar as they did not have power over and control of the social structures.
This study bridges the gap between both of these aspects, since it looks
at the creation of cultural and social practices.

Means of communication such as the telephone are not "fixed natural
objects ... They are constructed complexes of habits, beliefs, and procedures embedded in elaborate cultural codes of communication" (Marvin 1988, 8). They are also products of political and economic
conditions. As such, they involve both class and gender relations, and
are the product of a historical process. The historical conditions at the
basis of a process of communication comprise three main elements of
development: the technology itself – a product of technical characteristics influenced by the social conditions within which it developed; the
processes of production and consumption intrinsic to its operation; and
the social and cultural practices it generates. In an industrial capitalist
society, means of communication are developed primarily as a means
of circulation of capital. Thus, the development of a specific means of
communication, involving particular skills, is controlled by certain
classes, which limit its accessibility to certain groups and determine its
uses. Determined uses of a technology are based on the dominant ideology, which reproduces the existing class- and gender-oriented cultural
practices.

Siegelaub suggests that in an analysis of the production of a form of
communication such as a telephone system, the growth of certain types
of communication can be understood "as the dominant response at a
given moment conditioned by two interrelated requirements" (1979,
11). The first of these requirements is an economic and structural
demand, which implies that a specific form of communication is necessary in order for a given mode of production to maintain itself and
expand. The second, a political and ideological exigency, depends on
social antagonisms and active or passive political conflicts which help
to determine how and why different systems of communication are
developed, and how and why different types of communication evolve.

This refutes empiricist theories, which propose that "reality is just as
it appears," that all relationships between individuals and social groups
are transparent, and that exchanges between producers and consumers
are carried out under conditions of equality and objectivity. As A. Mattelart (1979) points out, the notion of class contradicts the "sacred
truth" that everyone is free to contribute to changing the patterns of
distribution and use of a communication system. The important feature
of a class analysis is that it unveils the notion of compulsion underlying
the imposition of a system of communication by particular classes upon

others. Situating an analysis of the telephone system within the social conditions of its development is the only way to unveil the elements of compulsion, as well as those of resistance, involved in the development of the system, and to go beyond the appearance of equality and objectivity that the system seems to create.

The presentation by some researchers of a "neutral," "objective," "universal" notion of communication without class and gender connotations (e.g., McLuhan 1964) has the effect of excluding alternative types of communication. An analysis of class and gender relations helps to uncover the various ways a communication technology could develop, and to explain why certain forms and systems of communication have become standard at the expense of alternative forms and systems.

To analyze a form of communication through social practice, and through the class and gender relations it involves, implies an investigation of how dominated groups, those without economic and political power, attempt to contest class and gender monopolization of the means of communication by powerful groups and to create other forms of communicative solidarity. Different classes may use different communicational forms, for instance, the family, social organizations, or solidarity among people in daily life, which way result in distinct uses of given communication systems (see Samuel 1975). To situate the creation of new communicational practices is to associate communication with culture. This "world of reference," for the dominated classes, is called "popular" culture by contrast with "dominant" culture, referring to the ruling classes (A. Mattelart 1983).

There is a dialectical relationship between popular and dominant cultures in which the latter appropriates practices of the former and, in turn, the former can reappropriate, through transformation and reinterpretation, some activities of the latter. Thus, popular culture exists as a culture of daily life based on contention and divergences. Since dominated groups are in a social position of non-power, their "ways of doing things" constitute the thousand micro-practices by which users counter the dominant control. This form of daily opposition to dominant institutions has no direct and immediate effect, but contributes to "changing things" not only in the immediate environment but also in the dominant and popular cultures (Willis and Corrigan 1983).

Popular culture, then, is a way for dominated groups collectively to translate their opposition into a strategy aimed at overwhelming the system and changing established structures. To counter this contention, the dominant apparatus functions to create cohesion in the ruling classes and to support the other classes in a way that will forestall the solidarity of emergent groups (Williams 1982). Thus, the relationship between changing means of communication and cultural change is problematic.

This relationship is even more complex for women. In the development of means of communication, women are in a situation in which they have little political and economic power *and* are subjected to patriarchal values supporting that situation. Women from different classes share these social conditions as a dominated group. Yet they may find themselves in different positions in terms of their actual access to a new means of communication, depending on the class to which they belong. A result is that while working-class women have limited access to the means of communication, ruling-class women have better access but are limited with regard to the types of use which those who control the means of communication allow them to make. This situation creates gender-oriented types of access to a new means of communication resulting from the social characteristics given to the means of communication by its developers. The technology, in turn, influences the forms of communication produced.

Forms of communication are also gender-oriented. Few researchers have looked at the way the communication process intersects with gender relations. The studies that have been conducted in this area are primarily concerned with images of women in television daytime serials or in women's magazines, and are limited to narrative structure and message. This type of model, says M. Mattelart, "hide[s] the very important fact that this ... product is the emanation of a particular system which ... crystalizes in its mode of organization the characteristics of its genealogy, as well as the role which it has been given in the production and reproduction of the social whole" (1985a, 19). We often forget that, in the confinement of their homes, women are responsive to means of communication that make their "exile more gentle, gratifying, in order to better reproduce that exile and effectively generate the conditions" of their isolation (M. Mattelart 1976, 296). Means of communication such as the telephone are sometimes the only link between housewives and the outside world. The telephone keeps them "in touch" with the family and the community. As such, it is influential in the development of some practices adopted by these women, since its use facilitates some of their activities.

It is important to note at the outset that the telephone is not merely a neutral technological instrument whose expansion has occurred naturally. It is a means of communication which determines the production and exchange of messages, and is related to the general structuring of the production and exchange of commodities in capitalist society. As such, its distribution creates a number of possibilities, and its use becomes a matter of interest to different social groups and classes. One means of control over access to a new technology is found in the degree of skill required to use it. Ruling classes might strengthen their control over a new means of communication if they feel that their own influence "depends on the maintenance of the social order," which could be disturbed by the new technology (Goody & Watt 1963, 314). Since the skills involved in the use of the telephone were minimal, issues of control centred on physical access to the instrument and on carefully planned development of its distribution and possible uses.

This book is concerned with the control exercised by the ruling classes over development of telephone systems in central Canada. For my purposes, a telephone system consists of a network whose telephone lines are connected to the exchanges owned by one company, and of the economic and political conditions that sustain its development and its specific uses. In Quebec and Ontario, several of these systems were gradually merged into a large system monopolized by Bell Telephone Co., through a multi-faceted process of political and economic interventions. Subscribers and independent companies used various means to oppose the tendency toward its control by private monopoly, sometimes forcing state intervention. Expansion of the telephone system was mediated by these oppositional interventions, though it was controlled by capital. At first, the telephone was intended to be used as a means of circulation of capital.[4] The system developed by Bell was designed exclusively for businessmen. This is not to say that capitalists decided the entire pattern of distribution and uses of the telephone system; on the contrary, groups who felt that their interests were not properly served by the service resisted and protested against the pressure toward monopoly. Women proved to be among the most active of these groups. Hence, this is not simply a study of "social control" by ruling classes, nor one of complete domination by male developers. It is, rather, an analysis of a dialectical relationship involving recurrent opposition to the planned expansion of the telephone system by Bell Telephone Co. This oppositional movement will be evident throughout the analysis.

The study is also concerned with the elaboration and solidification of practices related to the telephone, especially by women. It looks at the pragmatic aspects of use as products of the *form* of the telephone systems, and at the type of interaction they allow. The use of the telephone is taken for granted by most people today, who have not experienced a home without telephone service. However, the telephone affected late-Victorian society in a different way. On the one hand, two-way long-distance voice communication was a novelty, and many people remained skeptical about its use. On the other hand, people had experienced other types of long-distance communication and kept using the telephone in ways which applied to these other technologies and were often irrelevant to the telephone.

I am particularly concerned with the way users, especially women, were oriented toward what came to be standardized telephone uses. In spite of some opposition, telephone companies, supported by the ruling classes, succeeded in imposing most of the telephone practices that are with us today. As a result, the telephone is often seen as a natural means of communication, as an extension of the self. I argue that the adoption of certain specific telephone uses to the detriment of others was not the outcome of natural development, but of a telephone system shaped by a particular social group, and mediated by the resistance of another group, women, which sometimes led to unexpected uses.

Women users had limited resources with which to oppose and resist the proposed practices; therefore, the concept of their opposition and resistance is difficult to envision. However, I apply this concept to concrete situations related to the economic and political development of telephone systems, and posit that women used certain means to oppose the political, economic, and cultural forces of development imposed on them in order to change them. Cultural differences between classes and between genders may create divergent reactions to the development of a form of communication. Telephone companies promoted the telephone on the grounds that it would render life more pleasurable for their subscribers, and gave instructions to orient its use. Yet, in the face of some opposition, they generally continued expansion in a previously planned way, changing only minor features.

This points to the importance of examining the development of a telephone system in relation to women's activities and social practices. It indicates the necessity of studying the relationship between that system and women's practices in order to see whether the pattern of distribution and uses that the system was endowed with by the ruling classes sustained and encouraged new opportunities for women, or whether it merely reproduced their traditional activities, obliging them to adjust to the practices of other groups. It also suggests that one look at the impact of the telephone on women: did it have a liberating as opposed to an isolating effect, or any effect at all?

The purpose of this book is not to discover why women used the telephone extensively. Their isolation in the home, their role as agents for family errands, and their tendency to be more social than men (Fischer 1988b) were obvious reasons for them to use the new technology. This book examines how women's use of a specific technology influenced its development, along with the social and cultural practices already existing in the community. Women's influence was not limited to the use of the telephone, however; the production of telephone calls by operators was also instrumental in the development of a specific telephone system.

Since the occupation of telephone operator created by the telephone business rapidly became a "female ghetto," and since the users of the domestic network were mainly female, I emphasize the role of women in the development of a Canadian telephone system. The coming of the telephone changed both the nature of women's economic experience as operators and their community participation as users. These changes are examined and situated through a critical examination of the political and economic components of the telephone's development and their social impact.

Thus, this book is concerned with the social conditions within which the telephone system expanded. The technical characteristics of the telephone, examined in chapter 1, are part of these conditions. As I pointed out earlier, technologies have a valence – a limited range of potential uses. Nevertheless, within this range, there are numerous possibilities of development. Chapter one investigates the various ways in which the telephone might have developed technically, and how some elements were retained at the expense of others. The telephone was said to have limitless possibilities of use, and this chapter examines how it technically developed into a private form of interactive communication between two parties.

In telephonic communications, the mechanical characteristics of the technology had only a *limiting* effect. The technical development of the telephone systems was *oriented* toward a specific kind of expansion. Although it was seen as a "universal" means of communication, due to the low level of skill necessary for its use, the specific groups that controlled its expansion oriented its development toward a pattern of distribution that limited its accessibility. Chapter 2 looks at the economic and political conditions that sustained the development of the telephone, which led to a specific social distribution of telephones and to the creation of certain uses, not necessarily adapted to and accommodating all social groups. It analyzes the development of the new technology of the telephone into a "public utility," taking Siegelaub's two requirements into consideration. In contrast to the largely eulogistic treatment the telephone and its inventors have so far received, this chapter stresses the ever-present economic incentives behind Bell Telephone Co.'s decision-making process. Confidential letters and other documents reveal the company's consistent rejection of unprofitable types of development, despite the telephone company's status as a "public utility." The contents of managers' diaries demonstrate the political pressure exercised by the company on different levels of government so that it could enjoy legally unhampered development. This chapter also looks at pressure and opposition exercised by popular groups against Bell's monopolization of the development of the telephone system.

An essential element of the political economy of the telephone system was the labour process, and a crucial occupation within the labour force was that of telephone operator. Operators were mediators between the users and the telephone companies. As such, their work was closely related to both groups: they had to be productive enough to satisfy the companies and reasonably amenable to the subscribers. In chapters 3 and 4, I am concerned with changes in the labour process due to the development of new technology, in particular the impact of changing technology on the telephone operators, both in terms of their conditions of work and in terms of their attempts to organize. Chapter 3 analyzes the relationship of operators with the telephone company, starting with the process of feminization of the occupation forced on the company by its adoption of a particular type of communication technology, and describes five stages of transformation of the labour process. I discuss the increasing subjection of the operators to the technology and the different forms of resistance that they used to counter that process. In chapter 4, I examine the relationship between operators and consumers and their contribution to the making of a "telephone culture." Life histories of operators show that as they became increasingly subjected to the technology, their relations with subscribers became increasingly impersonal, so that their role shifted from one of "community worker" who provided a range of emergency and other social services to that of a detached "connecting voice."

Telephone operators were not the only agents of change in the development of the telephone. Users, especially women, also played a significant role, forcing the industry into accepting new and unexpected uses of the system. Chapters 5 and 6 examine subscribers' reactions to and participation in the development of the telephone, which resulted in telephonic activities that influenced the distribution of the telephone system as well as its use as a form of communication. In chapter 5, I compare women's communicative practices before and after the advent of the telephone, and demonstrate that today's taken-for-granted attitude toward the telephone was acquired, sometimes with difficulty, during the early period of its development, through various attempts to use it in a "rational" manner. In chapter 6, the pressures used by the companies to orient women's use of the telephone toward planned and expected practices, and women's unexpected responses to these pressures, are prescribed and analyzed. I reveal how this relationship between female subscribers and the companies, and between female subscribers, generated a particular culture of the telephone. Women drew the attention of the telephone industry to sociability as a use for the telephone, and thus indirectly helped not only to enlarge its expansion, but also to develop new cultural practices. This chapter documents the ways in which the development of a new communication technology can create a contradiction between privatization and socialization of women's communication. For example, the telephone, by providing instant connection, helped to widen women's opportunities for social contact, but limited their relations outside the domestic sphere.

CHAPTER ONE

"Freedom Can't Be Overheard"

In "Electrical Communication: Past, Present, and Future," F.B. Jewett, vice-president of American Telephone & Telegraph (AT & T), asserted that the technical aspects of the telephone presented "limitless possibilities of development" (1935, 178). He maintained that the technology was flexible and could expand in any direction. What factors influenced its development into the telephone system we have now – that is, a private-line system oriented primarily toward communication between two parties? Who controlled its pattern of expansion, and under what conditions?

Various possibilities for telephone use emerged in the course of the system's development into a monopoly. The telephone system in central Canada, which was constituted as a public service at the beginning of its existence, gradually came to be controlled by a private monopoly whose *raison d'être* was to make money. For subscribers, who were mostly businessmen in the early days, the incentive was to purchase a technology that would be profitable to them in terms of saving time in business transactions and lowering labour costs. The profit motive influenced both the form of the telephone system and the nature of access to it.

Thus, an analysis of the relationship between the political-economic development of a telephone system and its influence on women's practices entails an analysis of its technical development, which is partly responsible for its selected uses and its pattern of distribution. Various possible uses were gradually introduced through technological developments, but the telephone industry set policies aimed at pursuing only profitable kinds of expansion. The development of telephone systems was also oriented by state regulation and legislation. Although state intervention came relatively late – in contrast to the case of the railways, for instance – legislation, especially from 1906 onward, shaped expansion of the system in ways suggested by the telephone companies, particularly Bell Telephone Co.

In short, the telephone system did not expand "naturally," that is, with a will of its own, as McLuhan would put it. On the contrary, as a technological object the telephone necessarily has a social aspect, since it was developed for particular social purposes. The characteristics of the system resulted from the interaction between the specific social forces that controlled its development.

TRANSFORMING SPACE INTO TIME

Some researchers adopt a deterministic perspective, asserting that technological development is uncontrollable and that people have very little say in orienting it (e.g., Ellul 1964; McLuhan 1964). This view discounts the political and economic conditions in which the technology develops. However, it would be just as problematic to adopt the opposite position by asserting that technology is neutral and that its effects are entirely created by those who control and/or use it. Any technology has a valence: it can be used in a limited number of ways, among which only some will be retained by those who control its development. This was the case with the telephone. To understand its particular development, it is essential to examine briefly its technical possibilities.

In 1877, the *Revue et Gazette Musicale de Paris* gave the following definition of a new mode of communication called telephony: "Telephony, *as every one knows*, is that new process which permits the transmission of the human voice by means of electricity" (BCA, sf. in 1877a, my emphasis).[1] Notwithstanding this naive assumption of general knowledge about telephony, the simplicity of the definition corresponded to the simplicity of the new technology. The amazement produced by the new means of communication was not due to its intricacy, but rather to its potential for extending the human voice over several miles. The transmission of voice tones and characteristics was the important asset of this new device, since telegraphy already provided long-distance transmission of information.

Indeed, the telegraph had been in use for almost half a century. A system of telegraphy had developed which, especially in big cities, comprised a comparatively complex and sophisticated network. In fact, the technology of telegraphy had indirectly contributed to the invention of the telephone, as both were means of long-distance communication activated by electricity. However, despite their similarities, the two technologies were quite distinct.

To understand the distinctions among different transmission technologies, Williams' typology for the study of means of communication is useful. Williams (1982, 55) distinguishes three types of use and transformation of non-human material for communicative purposes: 1) "amplificative" means of communication, ranging from simple devices such as the megaphone to the most sophisticated technologies for direct transmissions; 2) "durative" means of communication, including such non-verbal communication as painting, as well as speech made durable by the comparatively recent development of sound recording; 3) "alternative" means of communication, which developed relatively early in human history, consisting of the use of physical objects as signs, then writing, graphics, and means of their reproduction.

This typology is based on the level of control over and access to a technology depending on the skills involved in its use. It centres on questions of social relationships and social order within the communication process. The two first types of communication can be differentiated from the third, since the skills involved in their processes are of a type already developed in primary social communication and imply no *a priori* social differentiation. This means that problems of relationships and social order centre on "issues of control of and access to the developed means of amplification and duration" – issues that are of direct interest to a ruling class. Some cultural practices related to forms of communication "can be given relative autonomy, within a monopolist social order, because they are already internally directed to the reproduction of this order, in its general terms, or internally directed at least not to contradict or challenge it" (Williams 1918a, 219). Nevertheless, for any class excluded from the development of these means of communication, they are easier to access than are alternative means, for which a primary skill (e.g., writing or reading) is crucial (Williams 1982, 55–6).

The problem of social order, however, is more than one of simple class differentiation. According to Williams, there is a reasonably direct and strong relationship between amplification and duration and the amounts of capital involved in their development and use. For instance, although it is relatively easy to establish a capitalist or state-capitalist monopoly on some systems of communication, direct access to the means of amplification and duration has created historical contradictions: it allows substantial flexibility and enables members of social groups and classes to be in contact with people beyond their own social system (Williams 1982, 56). In the process of development of means of communication, there sometimes exists an "evident asymmetry" between the social possibilities of a communication system and the institutions which could provide access to it. For instance, the fact that people in a particular place are provided with a communication system does not ensure their access to it. The possibilities of full access are determined by networks of institutional arrangements, and the "organisation of differential access to the most developed communications systems both corresponds to and [is] itself an integral part of social organisation." Hence, because a communication system and its technology sometimes stand in contradictory relation to the institutions that control it, it becomes, against their will, "a major organising force" in society (Williams 1981b, 228). However, limited access to these means makes them marginal in comparison to the huge, centralized, capitalist communication system (Williams 1982, 56). Community-television programming, for example, is relatively insignificant compared to programming by large networks.

For the study of development of the telephone system in relation to cultural forms, Williams' model offers an interesting alternative to the static and deterministic categorizations generally used (e.g., Brooks 1974, Latham 1975, Pool 1977, Singer 1981). It allows two levels of investigation. First, an investigation of *class differentiation in terms of access to and control of the telephone system* suggests that different opportunities for use of the telephone may have been created according to class. Second, an investigation of the *differentiation of access and control in terms of the availability of the telephone* indicated that availability is possible even when the telephone system is controlled by a capitalist monopoly.

The telephone can be categorized as an amplificatory means of communication since, technically, it can be used by anyone who is able to speak, whatever the language.[2] But if in principle anyone could use the telephone, in practice its use was limited by policies set by the telephone companies. Thus, an important issue in the development of the telephone system is the social impact of the selection of some of its numerous technical possibilities, to the detriment of others, by the social group that controlled that development.

As a means of interactive communication whose use did not require any specific skills, the telephone was symbolic of Victorian bourgeois and petty-bourgeois ideals of universalism. However, the control of its social development by private interests limited its universal expansion. The production of a telephone system in capitalist society tended to reproduce that society's class and gender differences. The availability of the telephone was based on two factors: its capacity for performing a certain type of communication at the expense of others, and its social distribution (discussed in chapter 2). Both factors were controlled by the ruling classes of the community in which it developed.

THE ROAD TO PRIVACY

It is not my intention, here, to write a detailed account of the technical development of a telephone system, since such a description would be very long and not proportionately interesting to most readers. Nonetheless, its social impact cannot be fully understood without a basic knowledge of some important steps in the development of telephonic technology. I will therefore briefly look at three components of a telephone system: the telephone apparatus, the wires and cables that connect the apparatuses, and the telephone exchange that transforms the individual connections into a system of interconnections.

The Telephone Apparatus

The early telephone apparatus involved three components: a system of transmission carrying a distinctive long-distance vocal communication; a system of signalling, indispensable for warning a subscriber that someone was "on the line" for the purpose of communication; and a system for discriminating between interlocutors, including a secrecy device which was supposed to keep the listeners at bay, thus achieving privacy in telephone calls.

The first instrument to be put on the market, in 1877, was the "box telephone." It resembled a wooden box camera with a hole, called the "mouth piece," carved in the centre which was intended to be used as both a receiver and a transmitter, so the user had to shift the box from mouth to ear. The transmission capacity of this apparatus came from a magnet whose power was proportional to its size. During 1877, the shape of the telephone changed dramatically, from an inconvenient wooden box to a "hand telephone" with a long goose-neck shape, small enough to be held in the hand easily (BCA, bb 1950, 93). It could be used as both a transmitter and a receiver, but most subscribers would buy one of each type, using the hand telephone as a receiver and fixing the box telephone to the wall to be used as a transmitter. This spontaneous use by subscribers was instrumental in the shaping of the telephonic commodity: most apparatuses, in subsequent years, were wall telephones incorporating both earlier types.

All of these apparatuses worked on the magnet principle of electric-power generation, and consequently were affected by the low power available. However, as early as 1878, the Blake transmitter, powered with a local battery, was developed. With this new transmitter, the different parts of a telephone set, sold separately until then, were combined into a compact type of telephone called the "white solid back." It consisted of three boxes mounted on a backboard with the magneto generator at the top, the Blake transmitter in the middle, and the battery at the bottom. The receiver was on the left side of the top box, and a crank on the right side of the bottom box was used to summon central, the office were all the lines met (BCA, qa, nda). The local battery powering the Blake transmitter was much stronger than a magnet, so it considerably increased the quality of transmission and was instrumental in speeding up expansion of the telephone system. However, it was a "wet" battery, and its acid leaked onto the wallpaper and floors of households that could afford a telephone (BCA, d 28012, 1880).

This development of the telephone apparatus had been the object of a complex legal process. The Blake transmitter was a modification of Edison's carbon transmitter, the patent to which was owned by Western Union Telegraph Co. in the United States. Legally, Bell Telephone Co. could not adopt a device derived from this patent without acquiring the right to use it. Bell secured Edison's patents in a successful lawsuit against Western Union. "Thus the introduction of the Edison carbon transmitter, with its threat of ruin to the Bell interests, was almost immediately answered by the Blake transmitter" (BCA, sf.in, nd, 5). According to John Radcliffe, of Bell Canada, "This really set the thing going. After the Edison's carbon, then proceeded a whole lot of improvements in the carbon transmitter, and Bell virtually came up with what they called the 'white solid back transmitter,' in service for years and years and years, and put the industry as a whole on its feet."[3] After this, the technical elements of the apparatus remained relatively stable even with the advent of the computerized communication system.

One aspect of the telephone that is often overlooked, but which represented, in the early days, "a drawback to the use of this wonderful instrument" was the signalling system (BCA, d 12015, 1877a, 49). The bell is an important part of the telephone apparatus: in a telephonic system, "no bell, no call" (BCA, d 12444, 1952). However, early telephones did not have a signalling system, and subscribers used a rather awkward method to signal their intention to communicate. When they wanted to attract the attention of another telephone owner, they either shouted loudly into the diaphragm or thumped on it with a lead pencil. The result was that telephone diaphragms were continually broken, until J.C. Watson[4] developed first a "thumper" and then a more sophisticated device called the "magneto generator," which, together with the "magneto bell," constituted the first ringing system. This system, however, had no discriminating device, and rang all of the subscribers' telephones at the same time. Lockwood described the consequences of the device: "A great increase in the number of electric bells manufactured ... [and] when that number was largely increased, the constant ringing of bells, melodious as it might be *per se*, and sweet as the Bells of Corneville, yet became a trifle monotonous, and once more the wearied ear yearned for rest and silence" (1893, 150–1). Hence, demand grew for a system of individual signalling. In a telephonic circuit of several connections, an individual signal operated in such a way that "*at the will of the operator* anyone of [the subscribers] ... [could] be called without signalling or attracting the attention of others (Lockwood 1893, 152, my emphasis).[5]

Thus, during the telephone's first years on the market, two components of the instrument – the transmission system and the signalling system – were developed rapidly so as to increase its use value. These components were directly related to the secrecy issue: the more privacy the technology could provide, the more useful it was for the class of subscribers targeted by Bell Telephone Co. Better transmission enabled better reception, diminishing the need to shout through the transmitter; an individualized signalling system eliminated the phenomenon of many people answering one phone call. However improved the transmitter was, though, it was still necessary to speak loudly to be heard at the other end of the connection: "Telephone users held the receiver like a time bomb ... shouted into the mouth piece at the top of their lungs, in fact, within six blocks, or ten if the wind was right they could be heard without benefit of the telephone at all" (BCA, qa, nd). Moreover, there was no way to prevent people from listening in on telephone conversations.

This lack of privacy was seen as problematic, and there were several attempts to develop a secrecy device to be attached to the telephone apparatus. As early as 1881, L.B. McFarlane, Bell Telephone Co.'s general manager for Eastern Canada, described a "secrecy switch" to be attached to the telephone instrument in order "to cut the line out" as soon as it was "tampered with by eavesdroppers" (BCA, sf.sec 1881). Several other devices for privacy were enthusiastically described in newspapers at the time. One of them, the "doo little switch," consisted of a button fixed on the telephone which, when pressed, rendered speech unintelligible for anyone on the line except the caller and the receiver (BCA, d 12015, 1882, 54). Later, the "listener trumpet" came on the market. It consisted of

a small trumpet made of vulcanized rubber ... [and] attached to the wall beside the instrument so that it may be easily connected with the transmitter and then, *at the will of the sender*, either moved back to its place or left in position for use. The great advantage of this device was that it concentrate[d] the tones of the voice, thus enabling the speaker to make even a whisper, inaudible to the bystander only a few feet away, distinctly heard at the end of the longest line. The trumpet render[ed] possible what ha[d] heretofore been impossible – the holding of confidential conversation by telephone. (BCA, d 30753, 1885, 6, my emphasis)

This device was considered a notable improvement in the domain of privacy. It had two functions: it prevented other subscribers from listening in on conversations and, because it increased the power of transmission, it allowed the caller to use a low tone of voice which could not be heard by people nearby. These two features represented a great advance over the preceding technology. The most interesting aspect of the device was that it was controlled by the callers. Thus, during the period when party lines were still the rule, the control of privacy on the lines passed from the will of the operators, who could control the number of subscribers to ring, to the will of the sender. The ability to make a confidential telephone call was now in the hands of subscribers who could afford the additional fee for a privacy device.

There were no further technical advances in telephonic privacy for almost thirty years. It is not clear, however, whether this was due to the success or the failure of the above-mentioned devices. All the same, in 1914, the onset of war made the issue of privacy imperative once more. A very sophisticated device was developed by an American inventor in that year. It could perform three functions: warning the users that someone was eavesdropping; identifying the eavesdropper; and, to a degree, measuring the length of the telephone call, since it could not be used for more than four minutes at a time ("To Stop Telephone-Eavesdropping" 1914, 733).[6] During the latter part of the war, a French captain developed another system to secure "military secrets" sent over the telephone. A device attached to the apparatus distorted the conversation at the point of origin and restored it at the point of reception. An inner mechanism made it impossible to interfere with the working of the device (Honoré 1919, 555).[7] "If a person at one end of the telephone wire desires to communicate a secret message to someone at the other end, with absolute certainty that neither intentional tapping of the wire nor their accidental crossing would enable anyone to pick up the conversation, he needs only to attach the new apparatus to the telephones which are in communication, and talk" (BCA, nct 1919b).

A last recorded attempt at securing telephone secrecy with a device attached to the instrument was made in 1917. This time, the apparatus was to be attached to the receiver, as opposed to the transmitter. It consisted of "a y-end tube … with ear-pieces on the ends of the y … employed to carry the sound to both ears at once, thereby doubling the volume of the sound." According to the reporting newspaper, "confusion of sound in telephoning" was proved to be due "to the fact that only one ear [was] used in listening" (BCA, ncm 1917d). This device, by raising the volume of transmission, was said to increase the degree of privacy.

Although some of these devices appear pointless to us today, the large number of attempts to increase telephonic privacy demonstrates the importance of the issue for users of the time. It is worth noting, however, that while all of the above-mentioned trials were made in order to develop a device to be attached to the telephone and controlled by the subscribers, the telephone companies (and particularly Bell Telephone Co.) saw the solution to confidentiality of telephone calls in privatization of the lines they controlled. This was a surprising conclusion for the companies to reach, since it was more expensive for them than allowing subscribers to buy their own secrecy devices. Had the telephone companies encouraged the development of secrecy devices attached to the telephone apparatus, privacy in the telephone system would have been entirely controlled by the users, and the structure of the system might have developed in a quite different manner.[8]

Wires and Cables

Once the telephone companies secured control of the telephone systems through the lines, advances in the technology of wire, which directly influenced the quality of the lines, was considered the most important step toward expansion of the business. While the telephone was in the amplificative stage, allowing limitless interactive communication, the wires constrained the form and extension of telephonic communications. As prime components of a telephone system, their quality affected three of its aspects: distance of transmission, quality of transmission, and privacy of communications. Because of their strategic role, research on wires received much time and energy in the telephone industry. The main problem was to find a way to increase cable capacity – that is, the number of calls sent simultaneously on one cable – and to extend the distance of speech transmission. The telephone industry achieved its goal through a three-stage development program: improvement in the quality of wires, the use of cables, and the development of devices that could be connected to cables to increase transmission signals.

In 1876, telephone calls were conducted on "open" iron wires which, at first, were organized into a "ground-return system": a single wire joined the two locations to be linked by telephonic communication. The wire "was connected with the ground, so that the earth provided the return path for the current" (BCA, sf.ind 3, 1938, 4). This system, which was very successful in telegraphy, proved to be a failure in telephony, since it was very dependent on atmospheric conditions and on other electrical currents in the vicinity. The directories of the 1880s warned subscribers that "[o]wing to atmospheric disturbances telephone talking is not always satisfactory, it being subject to considerable variation. At times, subcribers one to two hundred miles away can be heard with astonishing clearness; at other times owing to the above causes, great difficulty is experienced and it is found impossible to hear distinctly" (BCA, qa, nd).

These disturbances were known as "induction" in telephonic parlance, and were dreaded by both the telephone industry and subscribers. In fact, the telephone systems were considered "a comparative failure ... on account of the fatal induction" ("*Notes*" 1878, 46). Subscribers complained of "meaningless noises" on their lines: "There was sputtering and bubbling, jerking and rasping, whistling and screaming. There was the rustling of leaves, the croaking of frogs, the hissing of steam, and the flapping of birds' wings. There were clicks from telegraph wires, scraps of talk from other telephones, curious little squeals that were unlike any known sound. The lines running East and West were noisier than the lines running North and South. The night was noisier than the day, and at the ghostly hour of midnight, for what strange reasons no one knows, the babel was at its height" (BCA, sf.ty 1901–16, 88).

The sources of induction were numerous – electric-light wires and tramways among them (BCA, d 12016, 1892, 109). As well, variations in temperature damaged the wires and decreased transmission strength.

The development of the multi-wire lead-covered cable, made of seven or more rubber-insulated iron wires, remedied these problems. These cables were strung on crossarms attached either to roofs or to gigantic poles of forty to sixty feet in height. Some of the poles carried as many as twelve crossarms (BCA, ls 1883); "by 1890, the wires on poles had increased until in the centre of the city they literally darkened the streets" (BCA, ncm 1916a). The telephone industry was pressured by city dwellers and state agencies to find a solution to "this wire jungle."

Federal and provincial enabling acts permitted Bell Telephone Co. to "construct, erect, and maintain its line or lines of telephone along the sides of, and across or under any public highways, streets, bridges, water-courses or other such places ... " (BCA, d 1069, 1880). This section of the acts gave the company the right to expand the telephone system where it judged necessary, provided that it did "not interfere with the public right of travelling" in such places, although it required "the consent of the Municipal Council" to carry lines and poles higher that forty feet above the ground (BCA, d 1069, 1880). Granted such power, Bell started to erect poles and wires everywhere in cities, often not bothering to obtain the consent of municipalities. As well, the company often put its poles and wires on privately owned land without the consent of the property owners.

The result was a protest campaign in the mid-1880s, led by individual citizens and municipal councils and supported by newspapers, to limit Bell Telephone Co.'s "wild" expansion and oblige the company to create a more aesthetic means of transmitting telephonic communications. In Quebec City the matter reached the courts. A petition, signed by "ratepayers and municipal electors of the city of Quebec," was sent to the municipal council. It claimed that the telephone poles were not only a "nuisance," as they "disfigure[d] ... the streets by the erection of the bare monstrosities at all angles of elevation," but also a "danger" which "increase[d] in the exact proportion of the number of wires so used" (BCA, d 12016, 1881). Therefore, the petitioners requested that "the said Bell Telephone Co. be ordered to remove immediately the said telegraph poles[9] already planted and to desert from further obstructing that street with telegraph poles" (BCA, d 12016, 1880b).[10] In the following few months, legislatures in Quebec and Ontario enacted bills supporting the section of the federal act on the matter. Nevertheless, the battle over poles continued for many years. Some cities, such as Toronto, refused to have poles higher than forty feet (BCA, slo 4, 1889, 41). Others, such as Montreal, denied the company the right to dig up the streets in order to place underground cables (BCA, slo 3, 1888, 50). Finally, some property owners protested against the use of their gardens for the erection of telephone poles (BCA, d 6077, 1903).

These pressures led to the replacement of strung cables by underground cables "drawn into an iron pipe filled with oil and sealed at the end" (Perrine 1925, 117). Although this change improved service greatly, the iron wires remained unsatisfactory, and, in the late 1880s, telephone companies started to replace iron wires with copper wires (BCA, bp 1980, 7). Newspaper reports of the time were enthusiastic about the change, although it did not seem to eliminate the phenomenon of induction. Indeed, many years later, a telephone-company manager admitted that a conversation over a distance of 290 miles on that type of cable was "the equivalent of face-to-face conversation in a relatively quiet environment like an open field while standing about 135 feet apart" (BCA, bp 1980, 2).

Once the problem of poles was partly resolved in some cities by the installation of underground cables, the industry sought improvements in the quality and distance of transmission. As long as transmission problems persisted, it was impossible to acquire the desired degree of privacy in telephone calls. In the first years of the twentieth century, the "metallic circuit," in which two wires instead of one were used for each individual circuit, greatly improved the strength and quality of transmission. Then "iron core tyroidal type coils," a device used on long-distance lines, brought "the most dramatic increase in distance" (Perrine 1925, 119), producing transmission that was "remarkably free from cross talk" (Rhodes 1923a, 98). Finally, in 1913, the development of a "repeater" matched with "loaded cables" allowed for very satisfactory long-distance communication (Rhodes 1913a, 100), even cross-country, which was indispensable for conducting business transactions over the wires in privacy.

While the second stage in wire development aimed for improvement in transmission quality and distance, the third stage was concerned with increasing wire capacity. New devices were developed to obtain more communication channels without increasing the number of cables. In 1904, the "phantom circuit" permitted three telephonic conversations on four wires. Then the "carrier current system" allowed four to five telephone calls on one pair of wires (Perrine 1925, 122–9). This trend has culminated in the fibre-optic wire.

The Central Exchange

All of these technical advances to the telephonic apparatus, wires, and cables were necessary in order to improve the use value of the telephone and, thus, the number of customers. However, to transform this aggregation of telephone connections into a system of communication, one more step had to be taken: the establishment of central exchanges joining the discrete telephonic networks in a large system.

The concept of an exchange came very early in the development of the telephone. According to T.N. Vail, president of Bell Telephone Co. in the United States, it was "adapted from the connection of telegraph lines together at a central office to put different stations into direct communication with each other" (1913, 316). The technical instrument that connected all the stations was called the switchboard, and was handled by the operator.

The switchboard was a complex instrument, which underwent several transformations over the years. The first switchboard, the Gilliland, was entirely hand-operated, and had a capacity of fifty lines. The warning system consisted of a "red drop", which fell when a subscriber placed a call, making a faint noise to catch the operator's attention. The time the user waited to be answered partly depended on the zeal of the operators. With the Gilliland, subscribers' names were used to identify the connections to be made.

The Gilliland was used until the late 1880s, when the multiple-magneto switchboard appeared, with four times the capacity. This technological advance permitted so many subscribers to be linked up that telephone numbers had to be used to establish connections (BCA, d 10900, 1887b; ls, 1883). This apparently small change had an important social impact. The use of names for telephone communications had lent a "personal touch" to the system, since the operators knew all of the subscribers by name. When the use of telephone numbers became functionally necessary, some subscribers were so indignant about being identified by number that it took a few years for Bell Telephone Co. to force all of its subscribers to use them.

In 1900, the common-battery switchboard began to replace the magneto system. This meant that the leaky battery was removed from individual telephone apparatuses, and a more powerful battery in the switchboard office distributed electrical power to each station. Subscribers had "merely [to] lift the receiver [of the telephone] and small light signal[led] the operator" at the central exchange (BCA, sf.op. 1930). This system did not change very much over the years; the most important modification was its continually increasing capacity.

The next significant step came with the automatic switchboard. In Ontario and Quebec, this switchboard was adopted by Bell Telephone Co. in the early 1920s, although it was available and had been used in some cities in western Canada and in the United States since 1905 (BCA, ncm 1906c). This type of switchboard could handle the entire communication between two subscribers without any human intervention. According to a newspaper of the time, "it insure[d] *perfect privacy*" (BCA, nct 1920a, my emphasis).

Every new technical development contributed to the building of a telephone industry, although the advent of the exchange was considered, in the telephone business, to be the most indispensable (BCA, sf.ind 1929). Bell Telephone Co. constantly attempted to improve its product in order to increase its utility for the social group that could afford it: businessmen. The company directed the development of the telephone in such a way as to ensure better control over its system, and this had an impact on subscribers' access to that system. This "amplificative" means of communication, to use Williams' word, had an intrinsic quality of universality which was to be constrained by its political-economic development. This affected the social practices allowed by the telephone system, as it influenced its pattern of distribution and uses. However, the technical capacity of the telephone was only one aspect influencing its accessibility and control. More important were the economic and political conditions within which Bell's system developed, which oriented it toward specific uses.

CHAPTER TWO

Killing the Competition

The development of the telephone system was related to the expansion of capitalist production in Canadian society. Capitalist industries supported this technological development because it meant better control over and possible rationalization of labour, and a reduction in the time necessary to exchange commodities. The ever-increasing mass-productivity of the capitalist mode of production created the need for a more rapid and sophisticated communication technology. The telephone not only accelerated the reproduction of capital by facilitating commodity exchange, but also reduced labour expended on such things as messenger services. Thus, capitalists were particularly receptive to its use. The telephone was developed first as a commercial instrument by private companies, which skewed its social accessibility and the pattern of its development.

The most promising class of customers for the telephone business was small entrepreneurs. Competitive enterprises were particularly dependent on means of communication through which individual capitalists could get relevant information on the national or international market. Already telegraphic networks had considerably increased the speed of circulation of capital. However, not only did the telegraph involve several days' delay in the completion of commercial transactions, but its use also required mastery of relatively specialized skills.

The telephone represented an ideal way for communications to be made rapidly from person to person, without the intervention of a third party. In principle, this meant that telephonic communication was accessible to all. In practice, however, the telephone companies, and especially Bell Telephone Co., had other plans. Since it was business subscribers who would make the industry profitable, Bell concentrated its efforts on developing a telephone system that would satisfy the demands of capitalists for speed and privacy, at the expense of other dimensions of communication, such as collectivity and personalization, which did not correspond to the interests of these users.

This form of development was justified by the fact that technology in capitalist society is developed primarily for profit. The industry controlled the development of the telephone system in order to guarantee maximum return, and this in turn meant that the system developed first and foremost to sustain circulation of private capital. On the other hand, since the telephone was designated as a public service early in its development, some social groups opposed certain aspects of its development as a purely business-oriented system controlled by a private monopoly.

MAKING A BUSINESS OF THE TELEPHONE

The story of the development of the telephone, especially in Quebec and Ontario, is primarily the story of Bell Telephone Co. Although Bell was not the only company interested in creating and expanding a telephone system in these two provinces, it monopolized the telephone business in the most profitable areas, leaving small towns and rural areas to independent companies. This monopolization influenced the pattern of distribution of Bell's system, and involved legal fights against other telephone companies and, ultimately, state regulation in response to pressure from these companies and from dissatisfied subscribers.

The early extension of the telephone monopoly can be divided into three developmental stages: establishment, consolidation, and systematization. This categorization is based partly on improvement of the technology, and partly on the economic and political systems under which the telephone monopoly expanded.

Bell Telephone Co. was established in the first period, lasting from 1877 to 1889. The technology was still at an experimental level; its performance, touted as "wonderful" in newspaper advertisements of the time, actually caused frustration and despair for both promoters and customers. Telephone development began modestly, in 1877, with the leasing of telephones for use on private circuits that were often built by the subscribers themselves between two or more locations – for example, between branches of a company, or between a company's office and its owner's household. By 1879, several small, isolated telephone networks had been formed in big cities such as Montreal and Toronto. In that year, the first exchange in the business district of Montreal was opened by Montreal Telegraph Co., which used Edison's telephone, followed by one in Toronto, opened by Toronto Telephone Despatch Co.[1] A few months later, in Montreal, a second exchange was opened near the first by a rival company, Dominion Telegraph Co., using Bell's telephone. This firm was to become Bell Telephone Co. of Canada (BCA, sb 80013C, 3144–2, nd, 1–2).

In 1880, National Bell Telephone Co. of United States, which had taken over the Bell family's rights to the Canadian business, sent C.F. Sise, a Boston businessman, to Montreal to co-ordinate expansion of the Canadian business. Sise became the company's managing director after its incorporation later that year. As an important shareholder in Bell's Canadian patents, Sise had great ambitions for the Canadian telephone business, and wanted to monopolize the entire Canadian system. For Bell, this period was one of investment, with the primary goal of laying a foundation for a system that would eliminate competition from other telephone companies. Upon his arrival in Canada, Sise immediately started to buy out the most promising competitive companies, putting their managers on the board of Bell Telephone Co. (BCA, sf.ind 1945, 3–4). If the competitors resisted this approach, Bell's strategy was to "kill" them.

At the very beginning of the development period, Bell Telephone Co. managers expressed their conviction that it was "absolutely essential that the telephone service should be under the control of a single company," and that that company should be theirs (BCA, d 12016, 1880b, 25). "We occupy the field, we are entitled to it, and propose to hold it," said Sise during his first meeting as managing director with local managers.[2] To reach this objective, they decided "to meet" the opposition and *to completely kill it*" (BCA, d 26606, 1887, 43, my emphasis). Bell was ruthless in its drive for monopoly. According to C. Skinner, general manager of People's Telephone Co. in Sherbrooke, Quebec, tactics with regard to independent enterprises constituted "a catalog of unfair dealing … [consisting of] opposition in the legislature, manipulating municipality councils, obtaining exclusive privileges for telephone service and exclusive privileges to connect with railway stations, obtaining injunctions, securing protests, cutting rates, obstructing the building of telephone lines, tampering with lines and telephones, and employing agents to deceive and mislead subscribers" (BCA, d 6680, 1905).[3]

To "put instruments in free," or "at any prices," or "for nothing" was one of the "unfair dealings" used by Bell Telephone Co. to crush the competition (BCA, d 9710, 1888). Of course, reduced rates lasted only as long as the competition lasted, and full prices were restored immediately after the "killing." In a confidential letter to Smith, the local manager in Kingston, Sise summed up Bell's policy on competition: "The natural result of this [reduction] will be, as you of course understand, a restoration to the old price; but we are temporarily compelled to run the business at a loss in Montreal for self-preservation, it by no means follows that we must do it elsewhere, and we have no intention of so doing ... We are forced to do this in order to keep our subscribers, and also to keep the opposition from going elsewhere" (BCA, d 9710, 1888).[4]

Although it was fiercely competitive, Bell Telephone Co. had fewer than two hundred subscribers in Montreal and Toronto combined, provided with party-line service with eight to ten subscribers on each line. Despite this rudimentary system, the company was quite profitable.[5]

After its incorporation, the Canadian company enjoyed a solid financial foundation. Not only was its stock owned mostly by the American Bell Telephone Co. – Melville Bell, the inventor's father, had sold his Canadian rights to that company – but the latter had supplied the Canadian firm with funds for expansion,[7] and with telephone equipment developed in United States by the American company.[8] As well, most of the legal battles over ownership patents were fought in United States and paid for by the American firm.

Support from the American company, along with Bell's competitive practices, made it difficult for other Canadian telephone companies to survive. Small capitalists were left with marginal telephone systems in areas considered by Bell to be insufficiently profitable. Where profitable systems had been developed by others, competition always ended in the ruin of small capitalists, whose investments were either lost or appropriated by Bell.

Despite Bell's solid financial footing, its telephone system in central Canada expanded slowly – though steadily – especially in the big cities. In Toronto, for example, the number of subscribers grew from a little more than one hundred, in 1880, to 1,500, in 1886; it reached 28,000 in 1909 (BCA, sb 80041b, 3146–3, 1938, 1). The increase in telephone stations[9] connected up in that city between 1886 and 1890 averaged only 443 per year, or a little more than one telephone per day.[10] The small number of customers did not prevent the company from opening telephone exchanges in big cities, however. Although, initially, exchanges were opened wherever demand was sufficient, the company very quickly began to plan their locations. Sophisticated means of evaluating the market were developed in order to find the most profitable areas for installation of exchanges.

Bell's policy of maximizing profit also extended to construction of long-distance lines between exchanges, and between towns and cities. Although these long-term investments were usually built at a loss, they provided the basis for establishment of the company's monopoly. In effect, they represented the essential links for a national system, and by monopolizing them Bell put other telephone companies in a position of dependence on its long-distance lines if they wished to extend their networks from town to town. On the other hand, exchanges were opened only if the number of potential subscribers was sufficient to make them profitable, and if these customers could actually pay the rates sought by the company.[11] This, of course, considerably reduced the opportunity for working-class subscribers to obtain telephone service.

The opening of the Beach exchange in Toronto exemplifies this point. In 1902, the local manager of the Beach area canvassed the territory, urging people who wanted a telephone in their household to sign a list. At the time, the Beach area was inhabited primarily by farmers, working-class families, and, during the summer, a small group of bourgeois families. According to the manager, "it was an easy thing to get a large number of people to sign the petition" (BCA, sb 80141b, 3146–3, 1902). Clearly, this group of people thought that the telephone could be of some use to them. However, Bell charged the prohibitive price of one hundred dollars per year for the service (BCA, sb 80141b, 3146–3, 1900),[12] so that only a few of the signatories could afford it: "The people most interested found it impossible to get 15 to 20 persons to agree to pay subscription. There are no industries in the place ... As a community, the occupants of the houses are not well off" (BCA, sb 80141b, 3146–3, 1902). The company claimed that it "received no encouragement" (BCA, sb 80141b, 3146–3, 1903) from such uncertain prospects, and decided to offer summer telephone service to accommodate the bourgeois. A few years later, when the class of permanent residents changed, and the bourgeois and petty bourgeois established themselves in that quiet sector of Toronto, Bell built an exchange.

Working-class groups and farmers in the Beach area were aware of the usefulness of the telephone and were willing to pay a reasonable price to obtain the service – in contradiction to the company's assumption that working-class people did not need telephones and had no idea of the value of the service (BCA, d 17958, 1905). Bell made its decision to open an exchange not according to the needs of the majority, but rather with a view to "see[ing] a fair margin of profit" (BCA, d 29870, 1880). Moreover, if an exchange did not meet the company's expectation of profit, it was closed. The remaining subscribers were either transferred to another exchange, when possible, or lost telephone service altogether (BCA, slo 14, 1899, 15–16). This policy also applied to long-distance lines: when they turned out to be unprofitable, they were cut from the system.

In spite of Bell's efforts to satisfy the wealthy classes, initially these people were not overly attracted to the new technology. In fact, service was so poor, especially in terms of transmission and delay for calls, that there were almost as many subscribers asking to be disconnected as people requesting connection (BCA, d 7934, 1887)! Technological advances influenced the rate of expansion of the telephone business. The combination of the long-distance Blake transmitter, the metallic circuit, and the multiple-magneto switchboard was the catalyst for massive extension of the system. These apparatuses not only allowed for better transmission, but permitted the connection of a much larger number of stations in each exchange, thereby accelerating the opening of exchanges.[13] Montreal was a good example of this. The first exchange, the Main office, opened in 1880. Seven years later, the second office, Uptown, was built. The next exchanges came much more rapidly: East, 1888; South, 1890; Mount, 1898 (BCA, sb 80013c, 3144–2, 1887, 2). In July of 1900, Montreal had 8,120 stations. In spite of the small number of subscribers due to the high cost of the telephone, revenue was growing, and the business was making definite moves toward monopolization.[14]

During the consolidation stage, from 1890 to 1904, Bell Telephone Co. was categorized as a public utility. In 1902, Bell's charter was amended to oblige the company to provide whoever could afford to pay "lawful rates semi-annually in advance" and resided "within the city, town or village or other territory within which a general service is given" with telephone service (BCA, d 1069–4, 1902). The amendment contained at least two discriminatory components in terms of access. First, Bell's high telephone rates excluded low-income groups from access to telephone service. Second, groups of pioneers establishing new settlements outside of a territory with general telephone service were barred from a useful system of communication. Bell attempted to get the best of the amendment. On the one hand, it was good policy to claim that its telephone system had been declared a public service by the state. On the other hand, Bell was not prepared to sacrifice profit just because its product was now a public utility, and did not modify its rates.

The consolidation period also comprised major developments in the technical ability to increase speed an improve privacy, which attracted a larger number of subscribers. This was a period of continual, although not spectacular, growth. It was particularly marked by important growth in long-distance lines. While a large number of subscribers had been essential for the establishment of Bell Telephone Co., long-distance service was indispensable for its consolidation, for three reasons. First, long-distance service guaranteed Bell control over the telephone systems in the covered areas. Second, long-distance service gave the telephone the use value that could secure the only class of subscribers in which the company was interested at that time – businessmen. If it did not provide a link between various points of commercial transaction, the telephone did not have any value for these subscribers. Finally, long-distance lines paid tremendously well. Indeed, despite the company's claim of unprofitability, entries in Sise's log-book reveal huge increases in long-distance revenue.[15] This was the result of Bell's opportunism with respect to the erection of these lines: since long-distance lines between big cities were the most profitable, the company emphasized their development to the detriment of long-distance lines in rural areas. During the 1905 Mulock Select Committee hearings, Bell Telephone Co. of Canada was accused of neglecting unprofitable lines: "So far the small towns, villages and rural districts have hitherto been almost entirely left out of consideration by those upon whom the duty of supplying telephonic facilities has been delegated ... The Canadian rates [are] excessive due to the over capitalisation of the Bell Telephone Company of Canada" (BCA, ncm 1905b).

Correspondence between the superintendent and the district manager in Toronto supports this accusation. Although the cost of rural lines was very low (BCA, d 1127, 1897), Neilson, the superintendent, wrote Swinyard that "a new specification for farmers' party lines ha[d] been issued ... It [was] not intended to make a special effort to secure this class of business at present ... [The management] strongly recommended that no more grounded party lines be constructed" (BCA, d 1127, 1904).

These new specifications were approved by the vice-president, who "disliked very much" the idea of spending too much money on "such a route," since it was "not one of those we think will pay the best," although "it would seem good policy" (BCA, d 7966, 1892). At the Mulock Commission, Sise proclaimed that the company preferred to extend its system "in populous centres, rather than in rural districts" (BCA, ncm 1905c).[16] (This did not mean, however, that it provided service to the majority in urban areas.) Given that people in rural areas were the most remote from services such as doctors, fire department, police, and so on, and thus the most likely to find a telephone useful, one might have expected that a company that had been declared a public utility would start with development in these areas. Here, another contradiction in the development of a public service by a private monopoly springs up. Bell Telephone Co. produced to commodity only when, and as far as, it paid. Consequently, the people who were the most qualified for access to telephones in terms of "need" were deprived of the service.

During this period, Bell's long-distance system developed to the extent that, by 1898, it covered a large part of southern Ontario and the most densely inhabited and wealthiest areas of Quebec: the regions around Montreal, Quebec City, and the Eastern Townships (figure 1).[17] It established a solid basis for Bell Telephone Co.'s monopoly in these two

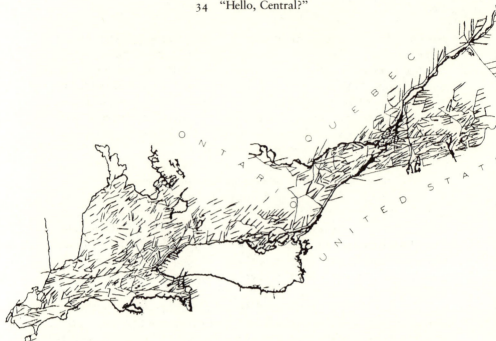

Figure 1: Bell's Long-distance Telephone System in Central Canada, 1898. (BCA, d 1544–2, 1898)

provinces. As Patten said, "The early erection of these lines consolidated the system to such an extent that competition was thereafter negligible" (1926, 90).[18]

The systematization period lasted from 1905 to 1920. There were no dramatic technological changes in these years, but rapid industrialization in the big Canadian cities directly influenced the growth of Bell Telephone Co. Telephone service expanded not only as a long-distance system, but within the cities as well. The latest technological innovations (common battery, underground cable, etc.) had improved telephonic communication, and the number of subscribers increased. On 1 January 1910, there were 12.0 inhabitants per telephone in Toronto, and 15.7 in Montreal (see table 1). These figures are deceptive, however, since they take into account the total number of stations, instead of the number of households with telephones. In fact, there were up to five phones in some bourgeois households, and some companies had even more, while the average number of connections in working-class households was near zero.[19]

During this period, Bell faced serious legal actions brought by municipalities and social groups. Agreements by major railroads to exclude

Table 1
Development of the Telephone in Selected Canadia Cities, 1910

City	Population	Telephones 1 Jan. 1910	Telephones per 100 inhabitants
Montreal	425,000	26,164	6.39
Toronto	340,000	28,310	8.32
Ottawa	80,000	6,570	8.21
Hamilton	68,000	4,196	6.17
London	52,000	3,438	6.61
Brantford	21,000	1,370	6.52
Peterborough	17,000	1,196	7.03

Source: BCA, tg 2(6) 1910, 2.

independent telephone companies from railway stations, franchises from municipalities that supported exclusively the development of Bell's system, and excessively high rates were challenged (Babe 1988, 18). Winning these actions institutionalized Bell as *the* telephone system in central Canada. However, commercial expansion required more than institutional recognition. Bell used various means over the years to promote its product.

"THE MARVEL OF THE CENTURY"

The style used in promotions of the telephone changed between 1876 and 1920. Although Bell Telephone Co. did not have a general promotional policy before the mid-1890s, management used various means to attract certain groups of subscribers. Ways of publicizing the telephone ranged from canvassing in areas that looked promising for development to advertising in newspapers, and included other, more or less orthodox means to increase the number of customers.

An important source of expansion of Bell's telephone system was canvassing areas in order to find potential subscribers. Early in 1878, Melville Bell selected agents to solicit subscribers for private lines all over Canada. In 1880, he sent out agents in two cities in Quebec – Montreal and Quebec City – and in twenty cities and towns in Ontario (Patten 1926, 60).[20] These agents – who generally became local managers if they found enough subscribers to open an exchange – were soliciting only businessmen and other men in the ruling classes.

The success of canvassing was interactively related to the extension of the long-distance telephone system: the opening of lines between large cities, combined with low rates and even free connections, was attracting new subscribers (BCA., sb 80910b, 3144–1, 1881b), while increased numbers of subscriptions in various towns and cities prompted erection of new long-distance lines (BCA, sf.me 1890). Canvassing was certainly one of the most important methods of promotion, since a direct contact with an agent of the company was thought to have more impact than any impersonal means of promotion. This kind of publicity, along with special rates and advertisements in newspapers, contributed to the opening of most of the exchanges during a large part of the early period of development.

Another means of promotion was through the press. The newspaper was deemed "a prompt means of reaching the reading public" (BCA, d 27344, 1887, 12). However, newspapers were not initially overly enthusiastic about the telephone. At the end of the 1870s, company management lamented the fact that newspapermen were devoting very little attention to telephonic experiments. In fact, it was this indifference that forced Bell Telephone Co. to adopt definite policies for publicizing its product in the press. Company management reached an agreement with some widely distributed papers that entailed the exchange of a free telephone in their offices for advertisements and insertion of notices at no charge. The company considered the deal a "profitable arrangement," since it guaranteed the "friendliness of the papers" and produced more numerous and more positive news reports on the company's activities (BCA, d 27344, 1887). In fact, managers, superintendents, and agents were advised, by Sise himself, to maintain "special relations with the Press" in their respective localities, because it could cause "a good deal of harm" to the company (BCA, d 26606, 1887, 24). To ensure the co-operation of the press, Bell invited groups of journalists to telephone demonstrations. Later, the company urged them to visit exchanges to see for themselves how "nicely treated" the operators were. Very few papers were critical of the monopolist enterprise. In general, the press endorsed the idea of a "natural" monopoly for a telephone system. Moreover, Bell made newspapermen aware of the superiority of the telephone over the telegraph in terms of speed and direct communication, two essential characteristics in news reporting. The result was that, very early, newspapers that could afford it started to use the telephone for most of their communications.[21] This turned out to be profitable for Bell. Promotion of the telephone in newspapers ranged from reports of exhibitions of the invention at the beginning of the period of commercialization to very sophisticated advertisements later on.

At first, advertising did not exist in its current form. The only effective way to publicize the telephone was via reports on experiments, which generally were conducted by A.G. Bell. Bell managers invited a carefully selected group of journalists to witness demonstrations of the utility of the telephone. Newspapers would then report on the advantageous characteristics of the "marvel of the century" (BCA, d 12016,

1879c). These demonstrations tended to portray the telephone as an entertainment technology more than as a means of circulation of capital. They continued until the telephone began to be developed on a modest commercial scale.

Bell Telephone Co. publicity also included various types of notices. These were written in a matter-of-fact style, merely describing the instrument, its functions, and its prices (BCA, d 12015, d 12016). After 1880, a new formula was featured, listing stores and other facilities with which subscribers could be put into telephonic communication, or names of the most prominent subscribers (BCA, sf.inv 1881, hit 1877–1909). It seems that these lists had two effects: the services offered by new subscribers such as grocers, druggists, lawyers, and doctors presented the telephone as a means for business transactions, and thus gave it an enlarged use value; and publication of the names of prominent subscribers appeared to attract other prominent customers. In any case, the publicity was aimed squarely at business class. It was only towards the end of the 1890s that advertisements tentatively began to emphasize the value of the telephone for housewives as well. Indeed, although residential telephones had been mentioned regularly in earlier advertisements, it was implied that their use was as a connection between the office and the household for the businessman. Starting in the early 1900s, advertisements were clearly directed at the housewives. The solicitation of female clients, however, stressed the necessity of the husband's consent for the acquisition of a telephone.

Toward the end of the first decade of the twentieth century, advertising of the telephone reached a turning point. It began to stress the characteristics of a new social order which could be obtained with a telephone, and was based on a system of beliefs that were relevant only to the wealthiest class. The telephone was presented as a "business saviour" (BCA, tdm 1882–1920). Sometimes the advertisements were addressed to wealthy housewives, though the telephone was rarely manifestly described as a means for person-to-person conversation. Rather, it was touted as a liberator of "women slaves," since one of its functions was to lighten the burden of household chores through "telephone buying," which involved shopping in "comfort" (BCA, ncm 1909c). It was a "guardian against nervous strain" (BCA, nca 1903–13), "a safety for your family" (BCA, nct 1911b), "a reducer of household fatigues" (BCA, nct 1911a) and a replacement for letter writing (BCA, nct 1912–13). In addition, advertisements claimed that the telephone maintained family ties and friendships and was even "the Implement" that could "save the Nation" (BCA, nct 1911b). In fact, the purchase of a telephone was presented as a moral obligation for a considerate husband and a good citizen.

Through these promotions, Bell Telephone Co. was implying that the telephone was "more than a product"; rather, it was "a way of understanding the world."[22] The basic premises of the system of beliefs behind expansion of the telephone were shaped by and for the corporate capitalist world – a primarily male world of wealth. On the one hand, the development of Bell's telephone system was constrained by the process of circulation of capital; on the other hand, the form it took helped to shape the type of interactive communication of the society in which it developed. Publicity was essentially directed toward the ruling classes. Only they, it was suggested, could be informed consumers, considerate husbands, and good citizens. Working-class groups, struggling in slums for food and decent shelter, were thought incapable of being telephone consumers, or even good citizens. From the company's perspective, they did not need the telephone and, in any case, could not understand its utility.

Prices for telephones set by company management, including a "reasonable" return to the shareholders – a group that included the managers – were well beyond the means of working-class families. A few skilled workers could, perhaps, by saving on other domestic articles, purchase a phone. In general, however, low-income groups were completely dismissed by the company as telephone consumers. Photographs accompanying advertisements illustrate that the pitch was directed toward the wealthy classes only.[23] But this situation narrowed possibilities for development, and soon a ceiling for sales of the telephone seemed inevitable. Thus, it was necessary to find new markets in order to sustain the growth of the company. A promising opening was the sale of extension telephones to wealthy families and businesses. The promotion of extension telephones forced the company to consider women's use of the telephone in its advertisements. For whom would an extension to a residential telephone be useful, if not women? Wealthy women represented the new market for sales. Although the price of an extension telephone was relatively low, bringing in only moderate profit, this market was preferred over the working classes.

Thus, although the telephone required *no* skills for use and was *theoretically* available to everyone who wanted it, economic conditions limited its accessibility to specific groups of subscribers. This supports Williams' assertion that control over and access to amplificative means of communication may be limited by its political-economic development. Despite the potential usefulness of instantaneous communication for all social groups, especially those who were socially or geographically isolated, control of the telephone system by private interests produced a particular class- and geographically specific distribution. However, as Williams suggests, the economic aspect of a means of communication is only partially responsible for how it develops. In the case of the telephone, its expansion was, at some point, tempered by political interventions.

REGULATING THE TELEPHONE SYSTEM

State regulation came rather late in the development of the telephone, and remained relatively mild until the mid-1910s. The most significant interventions were in terms of fixing of telephone rates, institutionalization of privacy on telephone lines, and development of the rural system.

Telephone rates were determined by the laws of profit of capitalist production. However, when the telephone system was defined as a public utility in the Bell's charter in 1902, not only did the company have to supply the product to everyone who wanted it and could afford it, but rates were now to be controlled by the state.

Telephone rates were first regulated in the 1890s, about fifteen years after the telephone's commercialization.[24] A small amendment to the company's charter in 1897 gave the Governor-in-Council the power to refuse rate increases to the company (BCA, ncm, 1897).

It is impossible to give an average price for the telephone for the establishment period, due to the wide variation in rates from user to user. For instance, of two Ontario lessees who rented telephones about two months apart, one paid twenty dollars more than the other (BCA, d 1065, d 8162, 1879). Added to the previously mentioned reasons for divergences in rates was a lack of organization within the company, partly due to the fact that the telephone apparatus was sold in separate parts, with different models sold at different prices. Rates also varied according to the type of connection – for example, residential as opposed to business stations. Moreover, the cost was lower if the subscribers built their own lines (BCA, d 24894-2, 1882).

In 1903, the government, through the personal initiative of Senator Lougheed, attempted to force Bell Telephone Co. to standardize its rates at the lowest level. Sise assured the senator that this would be done "at the earliest possible moment" – not at the lowest rates, "but possibly they would all be raised to the highest" (BCA, sb 59497.110, 4167, 1903). This attitude put pressure on the government to place Bell under the control of a government agency to fix telephone rates.

Formal, organized control over telephone rates by the Railway Commission began in 1906, under an amended Railway Act, to stop random fluctuations in telephone rates. Before this time, rates had been mainly controlled by the laws of competition.[25] Since the use value of each company's product was about equal, the price was determinant in the sale of telephones. Bell Telephone Co.'s ultimate aim was a Canadian monopoly, so it tended to cut prices to "kill" the competition, and this produced fluctuating rates. Local managers regularly reported consumers' attitudes to prices set by various companies to general management in Montreal, which adjusted the rates in relation to divergent situations and criteria created by competition or by other geographical and social elements. The Railway Act obliged the company to establish standard rates based on specific factors, and to ask the Commission's permission for "outstanding" increases. However, state regulations regarding rates were based on rather vague criteria.

In practice, government agencies exercised some authority over Bell Telephone Co., and had the power to control the amount of profit the company made. Each time Bell wanted to increase rates, it had to submit a financial report supporting the request to a special commission. Rate decisions were based on the concept of "a reasonable regular dividend," a loose and subjective criterion for the assessment of prices, all the more so given that, according to some observers of the time (BCA, t 6(3) 1903), the financial records presented by Bell to the commission were, if not falsified, at least "organized" to support its demands.[26]

The commission was formed in response to pressure from groups of consumers, and by some municipal councils, that felt that people's rights were being infringed by Bell Telephone Co. Its mandate was "to insure the equitable exercise and preservation of those rights." The very fact that Bell agreed to be supervised by the commission, however, suggests that the latter's effectiveness was questionable. Indeed, Bell management deemed creation of the regulating body to be not the result of abuse, but a preventive measure against possible abuses. Fixing rates remained entirely the responsibility of the company. "Until such utilities have made a sincere effort to co-operate in carrying out the purpose of such laws, they have no right to find fault with them," said a manager. Bell considered the Commission to be a "policeman" with regard to public utilities and their activities.[27] However, the Commission had no say in the company's management (BCA, d 3379, 1926–27, 7). The property of the public utility was private property, owned by a group of capitalists. Accordingly, the price of the telephone was fixed and regulated in order to allow for profit and accumulation of capital, while staying at a "reasonable" rate, affordable to a "reasonable" portion of the population.

Prices continued to vary either because of "discrimination" by the local manager in assessing rates to users (BCA, d 29144–53, d 29144–54, 1884), or because of special "considerations." Indeed, certain social groups were given free or reduced-rate connections: doctors received a ten-dollar-per-year discount (BCA, d 29915–4, nd). Table 2 lists the connections made free of charge in Montreal in the early 1880s.

The total of free telephone connections listed in table 2 is 146 – a fair number in a city with a total of less than one thousand telephones in use (BCA, sb 80013c, 3144–2, 1880, 6). Except for the "charitable" connections, all of them appeared to be advantageous to Bell. For example, some of them gave the company an opportunity to put telephones in public places, while others encouraged good relations with politicians at various levels of government.

The first public manifestation of dissatisfaction with telephone rates occurred in 1897, when Bell asked the Governor-in-Council for permission to increase its rates. The reason put forward by the company for the request was that maintenance and operation costs had increased by 30 per cent due to operation of the electric-railway and -light systems (BCA, ncm 1897), which had increased the level of induction over the wires and was so detrimental to the quality of transmission that Bell's system was not competitive any more. Thus, competition compelled replacement of the old instruments of production with new or modified ones before the expiration of their "natural life." The company was attempting to have subscribers pay for the resulting financial loss and for safeguarding its monopoly.

Although opinion was divided over the rate increase during the inquiry, there was agreement among politicians that the telephone system led to a "natural" monopoly. In response to Toronto alderman Scott's strong opposition to the proposed increase, which he considered "liable to add to the power of a monopoly," the Honourable A.C. Blair, of Ottawa, pointed out that "the telephone system must necessarily be more or less of a monopoly." City solicitor Fullerton supported Blair's argument, saying that when priced fell in the provinces because of competition small companies went bankrupt. Finally, Bell's representative summarized the debate: "[I]t was in the public interest that there should be but one telephone system in the country as it gave better service at lower rates than could be secured by a number of companies. The Bell Telephone Co. only sought the power to increase its rates in order to give a more efficient service. Without those the service should not be given" (BCA, ncm 1897). The subcommittee granted the company permission to increase its rates.

In 1918, a well-organized protest movement aimed to counter a new request for rate increases. The municipalities joined together to oppose the request, citing the company's failure to supply service upon application and to maintain the telephone system in good working condition despite the high prices it charged. In Montreal, La Presse, a working-class newspaper,[28] started a campaign against Bell Telephone Co.'s

Table 2
Free Connections in Montreal, 1914

Type of Connection	Number
Employees, pioneer (1), local managers (2)	15
Roof accommodation, or roof appraiser (1), or Tiffen estate (1)	33
Fire department and city	9
Charitable	20
Alderman, premier (1), director (3), president (1)	13
Contra.	4
Pole accommodation	1
CPR and Telephone Co. (1f)	5
Gazette	1
Pay stations	10
No reasons given	29
Architect	1
Honourable Mitchell	1
By order of Mr Sise	2
Road inspector and city treasurer	2

Source: BCA, sb 80910b, 3144–1, 1914.

monopoly, claiming that it was the cause of bad service and of high rates, which made telephone service inaccessible to working-class families. The average worker earned less than twenty dollars per week. Because of inflation, it cost the majority of working-class families an average of eighteen dollars per week to purchase basic food stuffs. According to Copp (1974, 71–3), a large percentage of working-class families in Montreal could afford only damp, overcrowed, and unsanitary dwellings. The cost of a telephone – twenty-five dollars a year for residences, then increased to thirty dollars – was prohibitive for these families (BCA, ncm 1918e). Nonetheless, according to Louis Racine, secretary of the St. Jean-Baptiste workers' union, and M.A. Duquette, general organizer for French-Canadian workers in the clothing industry, the telephone rate hike had indirect effects on workers' budgets. Small businesses, which were to be the most affected by the increase, were where working-class families bought their food and clothing. The increase would undoubtedly be passed on through prices of commodities, and thus would be absorbed by these families (BCA, ncm 1918g).

The population in general was skeptical about Bell's justifications for an across-the-board increase in rates. First, since the war was over, metal prices were expected to drop significantly, according to David Giroux, a well-known workers' leader in Montreal. Second, no one really believed that Bell was going to give a 75 per cent salary increase to its operators, which was one of its reasons for the rate hike. Finally, the tax increase cited by Bell affected everyone, and people could not understand why the company was asking subscribers and indirectly, non-subscribers to absorb this added cost (BCA, ncm 1918g). Bell's financial reports, published by *La Presse*, confirmed suspicions about the company's financial status. Net profits for the years 1913 to 1917 were as follows: $503,732 in 1913; $210,837 in 1914; $223,849 in 1915; $470,450 in 1916; $533,070 in 1917. During that time, the company did not increase its capital, which totalled $18 million in both 1913 and 1917, and its debentures were steady at $11,640,000, yielding additional interest (BCA, ncm 1918h). These figures had been supplied to *Moody's Directory* by the company itself *before* it decided to apply for a rate increase. Still, Bell wanted a 20 per cent increase in revenues.

Not surprisingly, workers interviewed by *La Presse* suggested that rates should be decreased, so that working-class families could afford the telephone: "[M]any workers and low-income people would be happy to have a telephone at home, which would be very useful for them, especially when they have to find a job. However, they are too poor to afford what is not a luxury, but a necessity" (BCA, ncm 1918d). These were the people whom Bell claimed did not want or need telephones.

La Presse's argument was that the telephone "ha[d] become a necessity in all spheres of human activity, and this [was] reason why its general distribution [was] essential, by means of popular prices. [This was a] serious injustice toward the working class" (BCA, ncm 1918i). Telephone rates ought to be "popularized," as were those for telegraph and postal services. According to the newspaper, only the state's "improvidence" was responsible for the inequity. Because of "obliging" political legislation, Bell enjoyed a monopoly in the telephone business which allowed it to serve its own interests before those of the "public." By permitting the company to exclude working-class groups from its group of subscribers via excessive rates, the state was failing to fulfill its role as "protector of the poor." Even more serious was the proposal that this exclusion be extended to public telephones as well. In effect, Bell Telephone Co. wanted to increase the cost of calls at public telephones in drugstores, and to decrease the druggists' share of the revenue by 40 per cent. As a result, the druggists adopted a resolution during a general meeting which forced the company to leave the druggists' share in the profit from pay telephones in their stores at fifty percent, or to remove the apparatuses (BCA, ncm 1918d). The removal of public telephones from drugstores would have deprived the working class of the only type of telephone service it could occasionally afford to use.

In its campaign against Bell, *La Presse* emphasized the failure of the state to make political decisions to modify the economic aspects of Bell's telephone system which permitted a tiny minority (owners and managers) to control the telephone activities of a vast majority. There was no regulation in Bell's charter – or in its amendments – that prevented the company from serving private before public interests. Further, although the state was pressured to control telephone rates, government agencies usually acquiesced to the company's demands.

A particular dynamic developed in the relationship between Bell Telephone Co. and the state in which the company repeatedly made requests for rate increases. On some occasions, the state was pressured by different social groups into a refusal. When a price increase was denied, however, Bell would be back before the commission a few months later with a slightly modified request. The commission would stress the impossibility of continual refusals and grant the increase. An analysis of press clippings from 1905 to 1920[29] documents this process. The company's requests for increases came so thick and fast that they sometimes seemed to fuse into a single request that lasted for years.[30]

BUILDING PRIVACY

One of Bell's rationales for increasing telephone rates was the need for capital to transform party lines into private lines, which were very much in demand among businessmen. The issue of privacy, which was to assume increasing importance over the years, became another opportunity for state intervention. A section of Bell's charter stated that "any person who shall willfully or maliciously ... intercept any message transmitted thereon, shall be guilty of misdemeanor" (BCA, d 1069, 1880). Some of Bell's activities were regulated by provincial enabling acts, which consisted of a short version of the federal act. The enabling acts covered elements under provincial jurisdiction, such as construction of lines on roads and streets and the power to acquire real estate (BCA, d 1069, 1881, 11–16; d 1069, 1882, 16). Telephone privacy was mentioned in the Quebec and Ontario provincial acts passed in the mid-1910s, which imposed fines of twenty-five dollars or thirty days' imprisonment for the offense of listening in on private conversations (BCA, sf.sec 1950).

The federal government's designation of Bell's telephone system as a public utility, as well as the act guaranteeing secrecy of telephone calls, presented serious problems for provincial legal systems. Bell Telephone Co. was obliged to provide an instrument to anyone who could pay, as long as the motive behind the request was lawful. However, the official reason for acquisition of the telephone sometimes differed from the real motive, leaving a loophole via which illegal businesses such as bookmakers, bootleggers, and brothels could acquire as many telephones as they could afford.[31] In fact, while "approved" businesses lamented the telephone's lack of privacy, "illegal" businesses thrived on the degree of privacy that telephonic communication gave them. Because it was difficult to prove the occurrence or content of telephone conversations, repressive forces such as the police could not find any tangible proof of their activities.[32] The problem was considered so serious that, in 1902, a Select Committee of the Ontario Legislative Assembly was "appointed to enquire into and report upon certain matters concerning administration of justice in the province of Ontario" (BCA, sf.il 1902). The recommendations of the committee affected several aspects of the telephone business. Bell Telephone Co. was forced to co-operate with law-enforcement agencies by allowing "peace officers" to search its records for signs of irregularities; if suspicious activities were uncovered, the company had to remove the telephone from the concerned premises immediately and disconnect the lines (BCA, sf.il 1902, 1–2).

These were the legal restrictions affecting the development of Bell Telephone Co. until 1905. The period from 1905 onward was one of greater political activity. In that year, the Mulock Select Committee of the Dominion on the Telephone System started an extensive inquiry into telephone expansion in general, and into Bell Telephone Co. in particular. This committee was empowered to examine every aspect of the telephone business in Canada, and to make recommendations to improve service. It was during the tenure of this committee that the issue of rural expansion of telephone service was raised. The committee had been created in response to pressure from the Union of Canadian Municipalities as well as from a large number of farmers and rural users, all of whom claimed that Bell's monopoly was detrimental to equitable expansion of telephone service, and that the company would be nationalized because Bell could not supply a useful system at a reasonable price, and refused to develop in remote areas. Indeed, in 1904, there were no more than 1,500 telephones in rural Ontario (BCA, d 3379, 1926, 6). L.B. McFarlane, Bell Telephone Co.'s manager for eastern Canada, "admitted that in the territory covered by Bell, the number of 'phones in use in rural communities represented only 1 in 1,246 of the population" (BCA, ncm 1905a).

Smythe has pointed out that Bell Telephone Co. of Canada limited its service to the largest towns of an area in such a way as "to skim the cream from large markets and to ignore the needs of the small towns and agricultural areas" (1981, 142). In fact, C.F. Sise declared to the Mulock Committee that "there [was] much better return from the expenditure of money on that work than there [would] be from expenditure of the same money on smaller lines."[33] The government's expert, F. Dagger, found that the company's service was unsatisfactory in terms of rates and of rural lines, and suggested that the state or municipalities would provide better service at lower rates. He also proposed that the federal government take over the intercity lines owned by Bell and put them under the control of the Post Office.[34] Local services could be licensed by provincial or municipal governments; "the licenses [were] to include the common law obligations of common carrier" (Smythe 1981, 143). Rural services would be provided by co-operatives and linked through the state service to larger towns. Dozens of witnesses from the Union of Canadian Municipalities and representatives from various rural areas supported Dagger's recommendations. The threat of state ownership was most serious, and the systematization or elimination of the telephone monopoly was at stake. If the threat materialized, the company would lose control over the structure of development of the telephone system in the country. Management resolved the crisis by using its political connections within governments and by manipulation of public opinion via demagogic publications stressing the "danger" of state ownership and the "marvels" of its own monopoly system. "Telephone Talks" were published by Bell Telephone Co. in several newspapers to convince the "public" of the indispensability of its monopoly (BCA, ncm 1905a).

Bell developed a policy of "good relations" with political figures which sometimes involved supplying free telephones. In 1900, for example, Sise instructed Dunstan, in Toronto, "on Scott's recommendation to furnish a free instrument to C.F. Chase Private Secretary to the Premier" (BCA, slo.15, 1900, 35). Early in 1901, another free telephone was placed in Senator (later Prime Minister) R.L. Borden's rooms in Ottawa. As well, jobs were provided to friends of government members. For example, in 1901, Senator F.D. Monk "called [Sise] to ask for a position for a friend of his" (BCA, slo.16, 1901, 13), which was immediately granted by Sise. Interestingly, these two senators, Borden and Monk, presented a petition on 27 March 1901, asking to postpone the reading of the bill regulating Bell Telephone Co. (BCA, slo.16, 1901, 30).[35]

Strategic political connections had been of major importance to Bell from its very inception. As early as 1877, before its incorporation, the company received patronage from some members of government. For example, the Minister of Public Works of Canada, A. Mackenzie, rented a telephone for forty-two dollars, "payable annually in advance, for communication specified to be between the office of the Minister of Public Works in the Departmental Buildings in the City of Ottawa, in the County of Carleton, in the Province of Ontario and Rideau Hall, in the County of Russell and Province of Ontario aforesaid" (Patten

1926, 24–5). A year later, a telephone line was constructed "connecting the residence of the Lieutenant-Governor of Quebec, at Sprucewood, with the Government office on St.Louis street" (BCA, d 12126, 1878). There was no price mentioned.

Political connections, then, were secured at both the federal and provincial levels of government. Although good political contacts at these levels were essential, management also thought it useful to establish ties with important people at other levels of government. In 1892, Sise instructed one of his inspectors "to make the acquaintance of some of the leading men at different points, so that you may be in position to obtain information regarding the service" (BCA, d 1277, 1892). In 1901, Sise wrote in his log-book that he had "recommende[d]" to his managers in Toronto "an amicable arrangement there [with members of the municipal council] in order to remove their opposition at Ottawa" (BCA, slo.16, 1901, 56). The opposition he mentioned was the joint action of the Union of Canadian Municipalities and rural representatives aimed at breaking the company's monopoly in Canada (BCA, ncm 1905b). These groups supported the Canadian Independent Telephone Association in its demand that Bell Telephone Co. establish a "telephone connection between every town and district in Canada" (BCA, nct 1916a).[36] The company considered political contacts to be an asset to the administration of its business, and did not hesitate to lobby when necessary.

These and other kinds of political connections proved to be essential during the tenure of the Mulock Select Committee. Reports and recommendations by credible witnesses were incriminating for the company, and most were in favour of the taking over of the telephone industry by the Canadian and provincial governments, or by municipalities. However, the company appeared to be able to apply stronger pressure than could the opposing groups. In fact, the committee never came to any conclusions. Mulock abandoned it to attend a conference in England and, on his return five months later, left the government on grounds of illness (Smythe 1981, 144).

Because the state presents itself as the representative of the collective interest, its agencies are sometimes forced to counteract the activities of private industry. In the case of Bell Telephone Co., the government usually took action against its internal policies when under social pressure. In giving priority to business development in big cities, Bell had very much neglected rural areas. Since the company had been granted a patent on the apparatus, it was difficult for independent companies to take over rural development. In 1885, in response to rural pressure, the Minister of Agriculture voided Bell's patent (BCA, d 12016, 1885a, 71). This, however, affected only a part of the technical apparatus[37] and, although competition became fiercer, Bell's monopoly was never seriously in jeopardy. Nonetheless, a number of independent companies, many of them started by doctors, were able to thrive in rural areas.[38]

Family practitioners had been slow to recognize the utility of the telephone in their practices, but, once they understood its value in emergencies, they started a movement for telephone expansion in areas neglected by Bell, especially in rural areas and small villages.[39] Their companies were often co-operative, and most of the time they comprised a small network, sometimes without an exchange. At any rate, these systems always consisted of collective lines, and housewives and their husbands were often said to hold or organize meetings on the lines. When there was an exchange, it was occasionally connected to cities through Bell's long-distance lines, for a fee. Ultimately, these small independent companies were all purchased by larger companies such as Bell. Several of them were merged with Bell between 1905 and 1910, when the company was forced to extend service to remote rural areas. Bell did not consider these companies to be up to its own technical standard, so it bought them, changed the apparatuses, and, of course, raised the rates. "They were very negligent about keeping their lines and instruments in repair," said McFarlane during the Select Committee inquiry, "and it resulted in their systems becoming unsatisfactory, consequently arrangements were made to take them in as regular subscribers for the Bell Company's local service in that district" (BCA, d 17958, 1905). To pay the Bell's rate turned to be impossible for many of these subscribers, who had paid little for the doctors' systems, and who often had paid in barter. Only in 1912 did hearings in Ottawa on Bell's activities result in the federal government ordering the company to co-operate with independent companies (BCA, nct 1913h). Meanwhile, independent subscribers had to pay prohibitive rates to be transferred to Bell's system and to use intercity lines.

Bell's progress toward monopolization of the telephone system also encountered other interference. Dissatisfaction arose among subscribers (BCA, 2(6) 1910: 1–2, tg 2(9) 1911: 4), as well as among newspapers.[40] The problem became even more acute during wartime. In a circular letter, marked "PRIVATE AND CONFIDENTIAL," sent to some superintendents and managers, McFarlane explained that "the uncertainty regarding future financial conditions in this county has forced upon us the need of instant curtailment in outlay, both of money and material, and whether on capital or operating account." Accordingly, all planned extensions were delayed and the canvassing stopped (BCA, d 24092, 1914). Shortages of materials worsened the quality of telephone service during the period up to 1920. In a long article published at the end of 1918, *La Presse* criticized Bell Telephone Co.'s monopoly, claiming that it was responsible for the bad quality and restricted production of the telephone: "[T]he biggest obstacle to the universal use of the telephone, to its legitimate development, to the use of new techniques of telephony, arises from the calculated apathy, the egotistical indifference ... of Bell's monopoly" (BCA, ncm 1918i). The newspaper stated that Bell's monopoly was "obstructive to progress" and aimed simply at "exploitation of the public" – at amassing ever-increasing profit without providing all of the advantages that should flow from a telephone system (BCA, ncm 1918d).

As the history of the telephone shows, the development of a new technology into a system is very much influenced by the political-economic and social conditions within which the system developed. In the case of the telephone, these elements were often in a contradictory relationship, causing resistance from some social groups. Since the telephone was developed for capitalist enterprises, many complaints came from those whose activities, they claimed, could not be carried on freely over the telephone unless telephone companies guaranteed complete privacy. Secrecy was essential to the mysterious transactions conducted in the harsh world of business competition, and was the symbol of "freedom."

On the other hand, the institutionalization of the telephone system as a "public utility" produced demands from working classes that rates be lower so that the system could be extended to all. Actually, the availability of this amplificative means of communication (to use Williams' terminology), which entailed a possibility of universality of use, was limited by capitalists' control of its expansion. Working-class groups challenged this mode of development, asserting that it led to a discriminatory distribution of the telephone. This tension between the obligation to provide privacy for the customers who paid the most and the obligation to extend service to the other classes as a public utility was present throughout the period studied. Women telephone users also influenced the development of the system, and this influence is examined in the following chapters.

CHAPTER THREE

The Making of the Perfect Operator

In its early stages, the telephone system required the involvement of operators as "mediating" components for the production of telephone calls. But telephone operators were placed in a paradoxical situation: they represented both a necessary element in and an obstacle to the production of instantaneous private interactive communication. Before the adoption of the automatic switchboard, they were essential to making connections between subscribers, but, as "human mediators" whose activities could delay or intrude on the privacy of telephone calls, they were obstacles to the development of the telephone service sought by the companies. The telephone companies attempted to produce operators with particular habits, skills, and attitudes. This led to the feminization of the operator labour force.

In early 1880, the first woman was hired by Bell Telephone Co. as an operator in the Main exchange in Montreal; soon after, company managers in Toronto also hired a woman. Thus began the replacement of an all-male operator labour force by one that was entirely female. At first, the company had hired young men who were telegraph messengers, thinking that since both systems were long-distance means of communication, the two occupations would be compatible. However, the two jobs required quite different skills. By 1880, the use of "boys" as operators was deemed to have been a total failure. A Bell Telephone Co. manager stated that if the company continued with boys as operators, "it was virtually facing bankruptcy" (BCA, ls, 1880).

In the 1980s, the issue of feminization occasioned a growing debate among social scientists, especially feminist writers. Much has been said about the feminization of teaching,[1] of office work,[2] and of domestic labour.[3] These models identify factors influencing the process of feminization of occupations, but none of them offers an adequate explanation. A "consumer choice" model suggests a weak concept of rational and conscious choice subjectively made by women, while the "reserve army of labour market" model, although it explains women's inferior position within the labour force in capitalist production, does not provide the elements that are essential to an understanding of the process through which women are hired in the first place. The "demand" and "segmentation" models give more concrete concepts (Lowe 1980). In the case of operators, the former model suggests that operating work became a female occupation because of changes made by the telephone industry in terms of job requirements. Sexually stereotyped requirements were assigned to the occupation, and with these came low wages. On the other hand, the "segmentation" model would emphasize that this new assignment of requirements was due to the fact that the operator's work was segmented from other occupations and did not require great skill. Moreover, turnover was not a major concern because it was easy to train new operators – training time corresponded to the time necessary to subject the operators to the switchboard.

In their analyses of the feminization of domestic labour, Briskin (1980) and Curtis (1980) argue that the issue of women's subordination is central to comprehension of the expansion of capitalist economy. Subordination is an essential element in the manipulation of women as domestic workers and as wage-labourers. Indeed, the concept of subordination is indispensable to an understanding not only of women's acceptance of second-rate and low-wage jobs, but also of their willingness to enter occupations whose demands include patience, obedience, and submissiveness. Women's subordination, institutionalized in the family and the work place, is central to the moral regulation of women in the public realm, where it works to encourage women to consider their work as a "labour of love." The occupation of telephone operator had many of these characteristics.

The degree of feminization of an occupation is affected by the state of the labour market and by the financial structure of the organization within which the process occurs. As well, traditional attitudes toward women in the society in which the process occurs influence the characteristics attached to the job.[4] Other factors, such as age, ethnicity, household status, job openings, and level of education also have to be taken into consideration to account for the timing of the phenomenon of feminization. Finally, there seems to be a general consensus among researchers that the feminization of an occupation results in a decrease of its social status and in a reduction in its pay.

The process of feminization of the opertor labour force in the telephone industry cannot be entirely explained using the assumptions suggested in the current debate on job feminization. Indeed, the telephone-company administrators who organized the hiring of women seemed to imply that the job had never been suitable to the male personality, although the work was directly related to the handling and repairing of technical apparatuses, a domain usually reserved for men.

How did the occupation of telephone operator come to be a female job ghetto? Women operators' contribution was central to the rapid development of the telephone industry in the late nineteenth and early twentieth centuries. Women's particular characteristics facilitated the transition of the telephone industry from a small-scale enterprise to a modern corporate capitalist monopoly. The early telephone system, consisting mostly of party lines, was like a public place in which people sought to have private conversation. To resolve the contradiction generated in a public system producing individual, private telephone calls, the telephone business developed a set of moral regulations to be applied to the operators. In fact, feminization of the occupation occurred during the late-Victorian period, in a context of intense moral regulation of women working in the public sphere. Operators, as mediators between subscribers, were told by the telephone companies to imitate the "moral values" conveyed by those with whom they were regularly in contact. Rules and regulations made by the company gradually covered not only the technical aspects of the operators' training but their moral education as well. In fact, feminization brought no decline to the job's social status – that is, its honour – but rather created moral barriers intended to improve the public "image" of the telephone system and, thus, to promote development of the enterprise. These factors motivated the telephone industry to shift very early toward women operators, despite the technical nature of the job, the "risky" night shifts, and patriarchal attitudes about women's work outside the home.

VICTORIAN MORALITY

At the beginning of its expansion, the telephone system was operated by a male labour force. As the industry developed, however, telephone companies started to hire women for operator work, asserting that their personalities were "better suited" to the work. By the end of the 1880s, almost all telephone operators were women. Indeed, the job of operator became a prototypical job ghetto from which the male labour force was almost entirely barred. Nonetheless, in the first years telephone companies, including Bell, had hired "boys" as operators. According to a male manager who had worked as an operator in 1878, "it was most natural to use boys as operators in the first telephone exchanges. Boys and young men had served as telegraph operators from the beginning. Then, too, it was unthinkable in the early eighties that a girl should be out after 10 o'clock so that the operating staff on the night shift was of necessity male" (BCA, sf.op, ndb).

In the mid-Victorian era, when the telephone was in its early stage of development, official discourse presented men as assertive and aggressive and women as submissive and passive. A woman was expected to be patient, tactful, and prudent, as well as forceful and courageous within her family in order to "use to advantage the limitless and indivisible affection she *owe[d]* her husband" (my emphasis). At the same time, she was required to be intelligent, discerning, firm, and energetic with her children to "properly direct their education." These social expectations for women were based on the distinct role to which each gender was subordinated. "A young lady, unlike a young man, is not in the least required to be present in the public sphere; on the contrary, it is in the interior of the family, under the eyes of her parents, that she ought to reveal all the treasures of purity, modesty, humility, and piety, which her heart possesses. These are the best qualities, the most beautiful ornaments of the young girl" (DMA, 9 (1882): 439).

The image of modest, pure, and submissive women was particularly widespread in the ruling classes, which attempted to convey their "ideal" moral virtues to the other classes through magazines and newspapers. An article entitled "La famille," in *Le monde illustré*, a newspaper written by philanthropically minded bourgeois and middle-class ladies and intended for the lower-middle and working classes,[5] made it clear that a woman "ought never to seek other than the pure pleasures which the interior of the family offers to her. Woman's life, that life full of love, self-denial, and sacrifice, should only be lived there; so that the obscurity of her surroundings might make her virtues shine forth more brightly" ("La famille" 1884, 31).

Furthermore, since a woman's place was in the interior of her family, she was not allowed to go outside alone, particularly not in the evening. According to the Catholic church, these "feeble creatures ought never to be left ... without effective control and supervision" (DMA, 13 (1901): 453). This constant supervision was to be performed by husbands, parents, or a responsible person, approved by the parents, to protect women against "a liberal current of ideas" and "pronounced tendencies towards excessive liberty" (DMA, 9 (1887: 66). This protective attitude applied to social rights as well. Because women had to be protected, their guardians needed the power necessary to subdue the "dangers." Inasmuch as men were to be the protectors, they required control of the means of protection. Since one of these was the material means provided by labour, men's power was extended to women's right to have a job. Hence, women's work was socially ill-regarded, and what was available was usually in the form of low-status, low-paid jobs. As the Archibishop of Montreal stated, "The ignorance of some and the ill-will of others work to produce an unfortunate situation in which people come to believe that equality [between men and women] involves identical rights, and women are urged to enter into a ridiculous and odious revolt with men on a field of combat where neither the conditions of the struggle nor the chances of success could possibly be equal. The realisation of such theories would be obnoxious to women and the family, and would shortly lead to the fall of one and the ruin of the other" (DMA, 14 (1909): 577).

That equal opportunity at work would cause family breakdown and women's "fall" was a common patriarchal discourse during the early period of development of the telephone industry, and one which guided its employment policies. Sangster stresses that Mackenzie King's "perceptions of the operators reflect[ed] a Victorian image of woman" (1978, 122), and that the Ontario press held that view as well. In press reports of the 1907 telephone operators' strike, "it was the moral, rather than the economic, question of woman labour which was emphasized" (Sangster 1978, 123). This Victorian view of women was not restricted to the ruling classes, however. According to Sangster, trade-unionists held similar opinions. "The views of many craft unionists were dominated by their belief that woman's role was primarily a maternal and domestic one." Moreover, "it was woman's contribution to the home rather than her status as a worker, which was most often stressed in the labour press. In fact, concern that woman's wage labour would destroy the family was very strong" (Sangster 1978, 126–7). It is not surprising, then, that the telephone industry recruited its first operators from a group of "boys" who knew a little about electrical communication, although they were by no means skilled labourers. Besides, young men were "cheaper" than mature and married men. Those hired were usually telegraph messengers, and were paid, very poorly, by the message.

A BOY'S WORK

In the early 1880s, when the telephone industry started hiring operators, the labour process was largely unstructured. It consisted mainly of improvised work using an unsophisticated technology which provided irregular production in terms of quality and quantity. The only requirements attached to the job of operator were that they be male and young: "They were boys 16 or 17 years old, and young imps, immune to all discipline" (BCA, sf.op 1930). Nevertheless, the telephone-company pioneers and managers had not "envisioned the fact, later proved by experience, that men and boys were temperamentally unsuited to the exacting duties of switchboard operation, and that this work was destined to be performed by members of the opposite sex" (Jewett 1936, 127). "Nature" had not provided "boys" with the right characteristics to be telephone operators. The larger the room in which they worked, the more the "boys" were undisciplined. John J. Carty (BCA, sf.op. nd[b]), an operator during this period, remembered that male operators were very "wild." William J. Clarke (BCA, sf.op, nd[b]) recalled the "fondness for clowning and practical jokes of the boys and young men employed in the operating room" when he was an operator. These jokes were "often at the expense of the subscriber." Moreover, male operators were not submissive; when "subscribers were rude the boys did not always turn the other cheek but matched insult for insult and curse for curse." Furthermore, according to Clarke, the boys were inclined to "wrestling on the floor," and failing to answer incoming calls. Their training was very casual. They would "drop into the office during a slack period on Saturday and Sunday to learn how to operate the switchboard at which [they were] to start on Monday." The majority of the male operators were doing operating work on a part-time basis and were employed in other capacities either within the telephone companies or in other industries. Nevertheless, some male operators asserted that there was the prospect of advancement within the telephone business, especially with Bell Telephone Co. after its incorporation in 1880, and that, since "most of them were ambitious," this "tended to keep them trying to do a good job in spite of the natural tendencies" (BCA, sf.op, nd).[6]

Even after Bell Telephone Co. started to hire women as day operators,[7] men continued to be hired during the day in order "to connect the calling lines with the trunk line to office called," or, in other words, to do the technical work necessary within the exchange. Besides, male operators continued to work on the evening and night shifts, some starting at 6:00 p.m., others at 10:00 p.m., according to the size of the exchange in which they were working. In big and busy offices, male operators were used as little as possible, and only during the times of day when business was slow. However, company managers concluded that "boys as operators were proved complete and consistent failures. They were noisy, rude, impatient and talked back to subscribers, played tricks with wires, and on one another" (BCA, sf.op 1930).[8] The company's response to such undisciplined behaviour was an attempt to enforce control over the male operators in order to respond to subscribers' complaints. In a letter to local managers in Montreal regarding a complaint he had received against a male operator, L.B. McFarlane assessed Bell's expectations for the operators' attitude toward subscribers: "He as well as other operators should know they have no business to talk back to a subscriber, and you can inform him that if we have any further complaint of this nature his services will be dispensed with" (BCA, d 27144-44, 1887).[9] Nonetheless, disciplining male operators did not transform their "unsuitable characteristics" into suitable ones.

THE "HELLO GIRL"

One particular aspect of the operator's work was that it involved both "masculine" and "feminine" tasks. It required some technical knowledge of electricity in order to cope with frequent repairs; as a mediating job, it entailed direct contact with subscribers, a sort of "labour of love."[10] It seems that, in 1876, the telephone business saw the technical aspect as more important, and hired young men for the work. Undoubtedly, patriarchal attitudes were primarily responsible for that initial decision. However, bad experiences with men operators convinced managers that, to keep their few customers, they had to emphasize the characteristics attached to the mediating aspect of the occupation. Thus, the position of operator came to be seen as "naturally" women's work.

When, in 1880, Bell Telephone Company hired the first female operator to work in the Main exchange in Montreal (BCA, sb 80013c.3144–2, 1880), early experiences in the industry had already pointed to a few specific requirements for operators which seemed essential to expansion of the telephone system. The telephone company concentrated on finding adequate operators who would accept low wages. The particularity of the product – an instantaneous telephone call – and the essential role of operators as mediators in its production, created a situation in which users and operators were constantly interacting. Since the subscribers were mostly businessmen, company managers decided that "courtesy" and "discipline" were the most important qualities to "perform successfully the duties of ... [such] subordinate places" as telephone operators (BCA, t 2 (6) 1901, 218). K.J. Dunstan, general manager for the Ontario district, insisted that "rules and discipline" were to be "strictly enforced" to ensure operators' courtesy toward users: "Mistakes may be overlooked, but lack of courtesy is an unpardonable offence. Subscribers will forgive a great deal if the operator is invariably pleasant and polite [and submissive], and the necessity of only employing educated and refined operators is very apparent" (BCA, d 1239, nd). The solution to the problem appeared to be the hiring of women as operators: "A woman would give better service and be a better agent," said C.F. Sise (BCA, d 26606, 1887, 9). It was suggested that young women be employed "at salary not greater than commission amounts to commission men." These men were telegraph messengers, hired as telephone operators, who treated the telephone as secondary to their own business and acted very independently from Bell management. In contrast, "the young women would be directly under the company's control, would attend promptly to calls and are as a rule more honest and careful than men" (BCA, d 27344, 1887, 3). Hence, the company's attempt to structure the labour force by starting a process of control over its male operators through enforcement of work discipline led to feminization of the occupation. The unresponsiveness of male operators to the process of subjection forced company management to adopt a policy of hiring female operators.

The first woman was hired soon after "a conference between Mr Forbes, president of the American company, Mr Sise, Mr Baker and myself [K.J. Dunstan] re girls as telephone operators. Mr Forbes said he believed they were used somewhere, did not know where and had heard nothing against them" (BCA, ls 1880).[11] Some time later, Mr Dunstan hired the "Misses Howell" as telephone operators for the Toronto exchange (BCA, d 2405, 1880). These women, and most female operators, came from the lower-middle and respectable working classes. They were "girls"[12] who needed to earn some money and, at the same time, could meet the moral standard of subscribers in the ruling classes. On the other hand, the companies also attempted to impose forms of comportment on subscribers, as a message in the telephone directories indicated: "Ladies are employed as operators; we ask for their courteous treatment" (BCA, td 1886). The occupation of telephone operator was a new opening for the large female reserve army of labour. An operator described the labour market for women at that time: "Few jobs were open to women then; even most stenographers were male" (BCA, sf.op 1930). Another operator of the 1880s added that "employment for women back in the '80s was limited to teaching school, factory work, restaurant and domestic work" (BCA, sf.op 1940).[13] Some women preferred being a telephone operator to being a teacher: "Teaching was prosaic, poorly paid and the profession was overcrowded" (BCA, sf.op 1930). It appears, though, that female teachers and telephone operators were from the same classes. Moreover, some of the requirements for getting the jobs were similar. For instance, for both occupations women had to present recommendations from three persons, including their clergymen (BCA, ls 1900, 41).[14] Age limits were also imposed, although Bell Telephone Co. accepted women as young as fourteen years of age as operators in the 1880s (BCA, sf.op 1940). This was particularly true when the hiring occurred through personal connections. For instance, a fifteen-year-old operator was hired in Brantford, Ontario, through the intervention of A.G. Bell himself. Her uncle, James Bishop, had a farm across from the Bell homestead in Brantford, and had a very

Increase woman position

special relationship with the inventor. "I can remember when Mr Bell used to come to visit his parents, [said this woman]. Often he would come over to my uncle's farm and help with the haying. He spoke of his telephone invention and my uncle said he would give him $1,000 if he could have a 'phone from his house in the city to the farm. Uncle James gave Mr Bell the $1,000 to help him in his research work" (BCA, sf.wo, nd, 18). A.G. Bell recommended the farmer's niece to the company's local manager in Brantford, and she was immediately hired.

When schools for operators started up, in 1900, the age limit was increased to seventeen. These women needed to have "good memories," "be tall enough so that their arms would reach all lines on the switchboard and be slim enough so that they could fit into the narrow spaces allotted to the #1 standard switchboard positions" (BCA, qa, nda). L.B. McFarlane declared, in the *Daily Witness*, that "in their selection the greatest care is exercised. They must be girls of irreproachable character, recommended by their clergymen" (BCA, d 12016, nd). These women could be only from "respectable" backgrounds.

Nonetheless, feminization of the occupation of operator created some difficulties for the company. Indeed, although the work itself required no particular skills, some technical knowledge in the field of electricity was useful. In the early days, the telephone system constantly required minor repairs and adjustments, which were effected by the "boys."[15] Since Bell managers thought women incapable of doing electrical work, for a few years after the hiring of the first woman operator boys continued to be hired to perform the technical aspect of the job. While women answered the subscribers, the young men connected the busy lines with the trunk. However, this was inconvenient, said a female operator at that time, as the boys played tricks on female operators whom they did not like (BCA, sf.op 1930). The problem was resolved by company managers, who provided all female operators with a "small book on telephone troubles and how to remove them" so that they could be "more self-reliant and ... fix small troubles that will come up" (BCA, d 27344, 1887). Thus, for the same wage, women provided the telephone industry with new labour power, supplying not only the "feminine" qualities of submission and courtesy, but the "masculine" characteristics of technical and mechanical skills as well. In spite of changes applied to the job definition and its labelling as a "female" occupation, some essential aspects of the work remained traditional male domains to which women had to adapt, at least until the end of the 1890s, when the improved technology eliminated the need for such interventions. It seems, then, that women had succeeded in adapting to the "male" characteristics of the work, while men had failed to adjust to its "female" features. For the telephone company, this represented a value which even its most chauvinistic managers could not ignore.[16]

The process of feminization of the operator's occupation did not happen instantly, however. An operator remarked, "Elimination of the masculine touch ... was not made with one gesture. It covered a long period of years" (BCA, sf.op 1930). The first hiring of a woman as night operator occurred in 1888 (BCA, sf.op 1930). Even so, male day operators were not eliminated all at once. In fact, women gradually replaced the "boys" as the latter were promoted to higher-wage jobs, quit, or were dismissed for bad conduct, or as new jobs were opened up.

Still, by the late 1880s, the job was seen as a female occupation which, as B. Lalonde, a female operator in Ottawa, put it, "requires a lot of devotion and brings very little gratitude from the public." The operators were "women who have *put their femininity to the service of the community*. Very few men would be patient enough to perform the duties of a telephone operator" (BCA, ls 1906, my emphasis). An Ontario newspaper, *The Watchman*, gave a very accurate summary of the sex-labelled characteristics defining operator work as a female job.

In the first place the clear feminine quality of voice suits best the delicate instrument. Then girls are usually more alert than boys, and always more patient.

Women are more sensitive, *more amenable to discipline*, far gentler and more forebearing than men ... Boys and men are less patient. They have always an element of fight in them. When spoken to roughly and rudely they are not going to give the soft answer. Not they. And every man is a crank when he gets on a phone. The personal equation stands for naught. He is looking into the blank wooden receiver and it doesn't inspire him with respectful politeness. (BCA, d 12016, 1898, my emphasis)

Apparently, then, woman's upbringing in Victorian society gave her all the necessary qualities to be a perfect operator "gifted" with "courtesy," "patience," and "skillful hands." In addition, she possessed a "good voice" and a "quick ear," and was "alert," "active," "even-tempered," "adaptable," and "amenable" (BCA, d 12016, 1898). As such, the female operator was considered "the heart of the place" (BCA, ncm 1908c), "the most valuable asset that a telephone company possesses," "the stock in trade" of that company (BCA, t 8 (2) 1904, 124).

Telephone-company managers did not hesitate to admit that the female operator was instrumental in the growth of the telephone industry. As L.I. McMahon, a Bell Telephone Co. official, pointed out, "If ever the rush of girls into the business world was a blessing it was when they took possession of the telephone exchanges" (BCA, ncm 1916b). Yet the female operator was very poorly paid, and her working conditions were difficult, in spite of the fact that Bell ameliorated her physical environment by adding rest rooms, comfortable cafeterias, modern bathrooms, and so on.[17] Indeed, as the telephone system underwent technical improvement, the task of operating became more exacting and the subscribers more demanding, so that only "girls with steady nerves and a phlegmatic constitution" could stand the constant pressure (BCA, d 12016, nd, 18).

In general, however, the operator's job was seen as a "respectable" occupation for women. Mary Rosetta Warren, an operator in Montreal from 1880 to 1891, clearly summarizes the "spirit" of the operators of that period: "I doubt whether the modern operator ever felt the thrill and glamour that we did in being part of the early telephone development ... It was a daily occurrence to be asked by a subscriber to say a few words to a gentleman who had never used a telephone before – which made me feel very important" (BCA, ls 1880–91).[18] In short, the company's decision to sex-label telephone operating as a female position did not decrease its economic and social status but, rather, increased the qualitative requirements of the job, and made it possible for the company to obtain better-educated and better-qualified applicants. The feminization of the operator position transformed a job which was considered at the outset to be a part-time male occupation into a "labour of love" for females. Although female operators were deemed to be finding their reward more in the "love" they received from the subscribers than in the wages they were given by the company, it gave women opportunities to be upwardly mobile within the operator force. In fact, while the dedicated operator was invaluable to the company, the occupation constituted one of the few opportunities for a "decent" job for young women, despite the low wages attached to it.

Although feminization of the occupation of operator brought new opportunities for women on the labour market, giving Bell the appearance of being a progressive employer, it is clear that the company started the process of feminization of its operating labour force in response to the acute problem of production which it had encountered with its male operators. After a few years of haphazard development, the telephone business started to expand more systematically, especially after Bell Telephone Co. was chartered in 1880. It is no coincidence that feminization of the operator occupation started with the incorporation of Bell. The high-quality work produced by women contributed to expansion of the telephone industry. Women, because of their upbringing, could be more efficiently exploited by the telephone industry, than could men.

However, contrary to what is suggested in the general de
inization, the operator occupation did not have a low soci
job was a very important part of the production process o
industry, and acquired a higher social status after being ..
"female" occupation, although its wages did not improve. This high.
status was because barriers of "respectability" were placed before pro-
spective employees, and because certain behavioural demands were
imposed on actual employees.

THE PERFECT OPERATOR

The behavioural demands made on women operators caused a transformation in the operator labour process, which was instrumental in
the construction of a specific structure of the telephone system. Operators responded to this transformation in a manner which was considered specific to women, and which was influential in attracting a larger
number of subscribers. However, feminization of the occupation of
operator did not eliminate all contradictions in the production of telephone calls. The creation of the "perfect operator," who would satisfy
the most demanding subscribers, was another source of conflict, which
also influenced the transformation of the labour process.

To examine these conflicts, it is instructive to recall Foucault's concept
of subjectification. Foucault is concerned with the mechanisms of discipline and control exercised over workers, either by external sources
of power or by self-repression and self-discipline. According to Foucault,
workers are "created" as subjects (that is, as *active* participants)
through three interrelated processes. The first is a process of objectification through training, drilling, application of rules, and the like, orig-
inating in a "mode of inquiry" established by the enterprise that
employs the workers. The second process is the application of such
dividing mechanisms as categorization of workers (skilled and
unskilled), also developed by the enterprise. In both cases, control comes
from outside the body of workers, who have no direct control over the
development of these practices. However, the third process is different.
Here, subjects are "self-creating" through their own internalization of
their identity as objects through the development of conscience and self-
knowledge (Foucault 1982, 777–8). Where this third process succeeds
(and it does not in all cases), control is then transformed into self-
control, whereas the two first processes bring the labourers to identify
themselves as subjects of the labour process. Hence, for reasons that
vary from one labour process to another, or even from one worker to
another, they come to identify subjectively with their position in the

labour process and to incorporate its characteristics into their consciousness.

Subjecting workers to the labour process implies a dynamic of power relations that includes control, discipline, obedience, and resistance. It involves forms of power that force individuals into repetitive actions which eventually turn into habits. It is exercised by control based on discipline within regulated and concerted systems of power. Subjectification involves an element of coercion, not only from external control, but also from internal discipline. It also entails a notion of "resistance" involved in the historical conditions influencing the process: the mechanisms of exploitation and domination have a complex relationship with the mechanism of resistance (Foucault 1982, 780–2).

The process of subjectification implies various forms of resistance against various forms of power and control. To understand these relationships, it is important to understand the forms of control and of struggle involved in the mechanisms of resistance. In the labour process, struggles are based partly on workers' refusal to be regarded as abstractions, as machines. They resist forms of power in their immediate, everyday life (not necessarily institutionalized power) which objectivize them by categorizing them, by imposing a "law of truth" that enables them to identify immediately what is true and what is false, and by marking their self-identities so that they recognize themselves as subjects. Thus, the resistance is against forms of power that make them subjects, either by control and dependence or by tying them to their own identities through a conscious internalization of their own subjection. These concepts are useful for understanding the contribution of women operators to the telephone industry, by helping to shed light on exploitation of the particular characteristics of the late-Victorian women which gave operators' work its value.

At the dawn of development of the telephone industry, the operator enjoyed almost complete autonomy in her work, as no one knew which particular skills and structure to apply to the labour process. She was subjected to no specific control, and the rules and regulations she followed were dictated by circumstances. In addition, she worked in poor physical conditions. Most of the time, she was alone – or with two or three others – in a small exchange in which she was expected to do all of the necessary work – not only producing telephone calls, but also maintaining equipment she was using and the physical environment in which she was working. She had "to light the stove each morning before beginning work" (BCA, sf.id 1977), and "to clean the place, sweep the floor and dust the material" (BCA, ls 1913); as well, she was expected to do minor repairs and keep her exchange in good order (BCA, d 26606, 1887, 10). In addition to these menial tasks, the operator acted as the company's "agent" (BCA, d 26606, 1887, 9). Her work was pivutal for the telephone business; courtesy with and consideration for the subscribers was thus part of it. She was meant to repress any desire to talk back (BCA, d 26606, 1887, 9), in accordance with her Victorian upbringing.

She was, as an operator recalled, a "jack-of-all trades" (BCA, qa, nda, 55). There was no clear definition of an operator's job. The telephone company expected that the operator would make the new technology work as well as possible. To fulfill this expectation, she had to cope with an indefinite working period which could vary from ten to sixteen hours per day. Since the working time was not specifically defined, "overtime work was not recognized by any compensating salary or time off" (BCA, sf.op 1940). In spite of this, the company counted on the operator's willingness to do overtime work when necessary. As C.F. Sise pointed out, during his first meeting with the local managers, "The operator ought to feel enough interest in the company to come after hours if there was anything to be done" (BCA, d 26606, 1887, 9). She was expected to be readily available at the demand of the company. This meant that "an operator had to be up and dressed" at eight o'clock every morning, go to a pay station, which was sometimes a "half mile from her home," and be "ready to go to work" (BCA, ls 1899–1911).

The physical environment in which the operator worked reflected the lack of definition of her task and work period. The technical apparatus was very delicate and subject to disruptions from snowstorms, street cars, and so on, which decreased the quality of the communication when they did not eliminate it altogether. The operator therefore had to shout into the transmitter all day, in order to be heard by the users. This proved to be detrimental to her health. In addition, the operator was prey to such dangers as lighting hitting the telephone lines, which often inflicted serious injury. Moreover, the rooms in which the exchanges were installed were very primitively furnished and badly heated: "Heating arrangements were very unsatisfactory, no heat until November 1st at times our fingers were numb" (BCA, ls 1888–96). There were no proper accommodations provided for cooking or eating lunch: "The lunch room was provided with a gas jet on table for heating water and a discarded opening chair or so ... On a wet day the room was rather odorous" (BCA, ls 1888–96). As well, the premises were often infested with vermin: "[R]ats ... ran through the office. For instance, when lunch time came, we had a lunch provided the rats left some for us. At night, the operators found it necessary to sleep on a table, and even at that the rats often carried away some of their wearing apparel" (BCA, qa, nda, 39). The company's desire to save on fixed capital had a direct effect on the operators' working conditions: the deprivation they suffered due to the physical environment and low wages exposed them to a degree of misery. This exploitation was considered necessary by Bell in order to support the development of a still precarious industry.

Yet what the operator lost in comfort, definition of tasks, and limits to work period was compensated for in her level of autonomy. The operator was only occasionally submitted to the direct control of the local manager, who spent most of his time away from the office in order to canvass areas surrounding the exchange to find new subscribers. Still, the managers did attempt to influence the operators: he installed in them the spirit of duty and loyalty that were said to be the qualities a good employee had to demonstrate toward her employer. As one operator said, "You know, we hear a lot about the 'Spirit of Service' of the telephone operator. She has it all right, and she had it in those early days. But I think the loyalty of the operator was reflected, as a mirror reflects light. It really came from the men who managed those central offices. You just couldn't be anything but loyal to your job and to the public when you were working for – or rather working with – men of that sort" (BCA, d 19406, nd, 11).

What this operator did not say, though, was that the financial incentive, so high for the local manager, did not exist for the operators. Their wages were not sufficient even to provide them with personal independence (BCA, ncm 1907b), and they certainly could not afford shares in the telephone company, which managers commonly received.[19] Moreover, there was no possibility of promotion at that time, since management positions were not open to women, and since there was no formal hierarchy among the operators themselves. The "spirit of service" developed from their relationship with men who were economically interested in the development of the company. This desire to copy the local managers' loyalty to the company was reinforced by "the close interest [that the local managers] maintained … in all of the operators' … personal welfare" (BCA, d 19406, nd, 11). In effect, the paternalistic relationship existing within the family was reproduced at the exchange office in the form of the manager's "personal interest" toward the operators, and constituted the moral control to which the latter were subjected. Their response was a willingness to be good employees. Accordingly, a good operator was one whose behaviour was considered to be the most appropriate response to the manager's attention: doing all of the tasks necessary to production of telephone service as well as willingly enduring all inconveniences. These loosely defined duties were based on informal rules set mainly by the local manager and suggested through his example rather than his instructions.

In the early 1880s, the moral issue related to privacy had not yet been seriously raised for operators; very little energy was spent on preserving the private aspect of telephone calls. It was expected, however, that operators would refrain from eavesdropping, although they were not clearly instructed to do so. As well, the use of telephone lines by the operators had not been yet restricted by the company: "When there was very little to do ... we would talk with [the subscribers] on the line. *There was no rule against it*" (BCA ls 1880–95, my emphasis). In short, in the early period, the company's efforts were directed toward production of a personalized telephone service.

Formal rules and regulations subjecting the operators to their work were instituted toward the end of the 1880s, after the telephone companies realized that some aspects of the work involved human activities that could be dominated and exploited. Management began to analyse various elements of the occupation as part of its search for more profit. The main objective of the telephone industry was to make the telephone profitable for the industry itself and for the subscribers, who were mostly businessmen. To attain this objective, the industry had to structure the labour process in such a way as to increase profit by acquiring maximum control over the labour force. It was assumed by management that segmentation of all aspects of the operator work would reduce the risk of human error (e.g., bad connection, insubordination, etc.).

To satisfy subscribers, who were assumed to expect a certain degree of privacy when using the public telephone system, the subjection of the operators was directed not only toward an increase of production as such, but also toward attitudes that met the "general," "ideal" values of the ruling classes. This was reflected in the company's rules and regulations, which covered not only the technical aspects of the worker's training, but also her "moral education," in order to develop a labour force of perfect operators.

It is impossible to examine separately the two aspects of the external power imposed on the operators: the formal rules, task definition, and training period were causes and consequences of the creation of job categories, degrees of hierarchy, and the like. The telephone industry was not different from other industrial sectors in this regard, and both components of the process developed in tandem.

As the technology advanced, the specifications and limitations of the operator's job became better defined, and clearer external controls began to appear. The period between 1884 and 1891 involved the formation of a hierarchy within the operator labour force, accompanied by more direct control and formal discipline. The first chief operator was appointed in 1884. She was put in charge of all of the operators (BCA, d 24096, 1884).[20] As a consequence, the relationship between the regular operators and the local manager became indirect. Operators' grievances and demands were mediated by the chief operator, who had the power to make most of the decisions concerning the workers. This also constituted a change in the gendered character of power relations.

However, the mechanisms of control over the operators were still relatively imprecise, as the definition of the chief operator's authority was loose and her means of control over operators' activities was unsophisticated. The result was a personalization of the duties attached to the position, and a style of supervision which varied with the person who had the responsibility. This was reinforced by the fact that the chief operator's duties were not defined bureaucratically, and were not distinctly known by the regular operators. In no way, then, could operators ascertain whether the control exercised over them was overly zealous or in accordance with company policy.

The main duty of the chief operator was to control the work of the regular operators and to check whether they were following the rules. To fulfil this task, she paced continually behind the regular operators sitting at the switchboard. She could not check all of them at the same time and could not hear what was said on the lines, but the operators knew that at any time the "chief" might come and check their work. Operators who disobeyed the rules had to be reported to the company and could be penalized, according to seriousness of the infraction, by a warning, a denial of the rest period, a stoppage of pay, a suspension for a few days or weeks, or utimately firing (BCA, ls 1884–1891). The application of the chief operator's duties meant some limitation on the operator's labour, and a decrease in her autonomy. Although they did not touch upon all aspects of her work, they included rules that constrained the operator's actions: there were now a proper way to answer the subscribers, time sheets to fill out, a specific tone of voice to use, a ban on conversations with subscribers (BCA, d 24096, 1884). Thus, the creation of the chief-operator position not only modified the regular operator's relationship with the local manager, but made her relationship with subscribers more impersonal.

This subjection to a "model" behaviour, which had been unknown to the operators during the period of personalization, suggested that these issues were such a barrier to profit that the company felt it necessary to structure telephone calls. The creation of a hierarchy within the operator labour force by the appointment of a chief operator was the company's first step toward a much more rigid structuring of the operator labour process, and toward regulating the mediating role of the operator in the telephone call. While it was still too early to speak of the making of a perfect operator, the subjection of the operators to definite rules and regulations was nonetheless underway via informal training. The result was that several aspects of the operator's work could be controlled by the company, and infringements of the defined rules could be punished.

These rules, while more precisely defined by 1890, applied only to the public aspect of the telephone system. They were enacted in order to better the service – namely, to improve the operators' level of productivity – and increase the company's profitability. The moral issue of privacy had not been raised; although the operator was expected to be considerate, the question of moral education or etiquette was not formally addressed. Perhaps this was due to the fact that the technology itself obliged the operators to intercept the conversation from time to time in order to see if the parties were still in communication (BCA, bp 1980, 3). This was undoubtedly an encroachment on private communication, although no one seemed particularly alarmed by the intrusion. With the advent of switchboard technology, the physical working conditions were modified. For instance, the operator had to wear a "harness headset" weighing more than six pounds during all of her working hours. This meant that she could work with both hands, whereas previously she had held the receiver in one hand and connected the calls with the other (BCA, sf.op 1940). However, although control over regular operators was tightened by the creation of a hierarchy, it was still relaxed enough to leave some room for short conversations with subscribers, slowdowns in the production of telephone calls, and unauthorized uses of the telephone lines. These mild forms of resistance were in response to the first attempt by Bell Telephone Co. at some type of bureaucratic structuring of the occupation of operator as a means of control over her labour. The imposition of control by means of constant supervision was resented by the operators, who, until then, had been mostly on their own.

FROM PERSON TO MACHINE

The next period of development, 1892–1901, saw a multiplication of the mechanisms of control over the operators, including official training by qualified staff and publication of general rules. The new training concerned aspects of the operator's work which had been neglected until then: the quality of voice and the speech to be used, and the moral values to be applied in the operator's relationship with subscribers. Moreover, official regulations connected the operator's performance to the outcome of her work. The purpose was to make each operator individually responsible for her production and to force her to become more conscious of her work and of her role. The company sought operators who would internalize and identify with the desired characteris-

Faucault

tics. However, this aim was not reached without opposition, as the operators' consciousness worked both ways: it encouraged self-aware-ness, as well as the resistance to exploitation.

During the summer of 1892, a pamphlet entitled *Rules and Instruc-tions for Operators*, signed by L.B. McFarlane, was printed by Bell Telephone Co. and sent to its operators. This pamphlet, subtitled "Oper-ating Department Rules," prescribed twenty-seven "duties" for regular operators and fourteen supplementary instructions for "toll line oper-ators" (BCA, d 920, 1892, 2). Some of the listed duties confirmed or reinforced those specified during the previous period. However, some new rules were added. Several duties concerned technical details, and gave precise instruction on how to handle the work in order to improve use of the telephone system. In addition, the instructions extended to the speech to be used in telephonic communications by providing stan-dard phrases to be employed on specific occasions: "When calling on signal lines, be careful to ring slowly and distinctly, and give the correct signal: then wait a reasonable time and if no answer, repeat once, if no response, notify caller, 'they do not answer'" (BCA, d 920, 1892, 2). *A new limitation was thereby imposed on the operator in relation to her language*. Situations in which she could not apply the standard responses were to be dealt with by the chief operator or her assistant at once, for, in order to enhance her control over the regular operators, the chief operator had by now been mechanically connected to each worker and could answer the "problematic" calls – when a caller "insisted on talking" to the operator; when a subscriber wanted "to retrace a call"; when a user called "by name only," instead of using the number, and so on (BCA, d 920, 1892, 2).

This standardization of the operator's labour was the first real step toward depersonalization of the relationship between operators and subscribers. The standard answers used hundreds of times every day came to be given in a tone which discouraged any personal contact. Operators internalized these tones of voice and phrases by constantly repeating them. The company was slowly achieving its objective. The more mechanical the work of the operator, the larger the number of telephone calls connected during her work day and, thus, the bigger the profit.[21] Decreasing the autonomy of the operator reduced the margin of error and the time wasted in connecting calls. Moreover, not only did the rules reduce the autonomy of the operators, they held them "responsible for errors, or slow and unsatisfactory service at [their] section while on duty" (BCA, d 920, 1892, 3). *This specification effected a significant change in the labour force by shifting the worker's respon-sibility from a social to an individual liability.* For the first time, the word "responsibility" was used by Bell Telephone Co. in relation to its operators. This suggests that the company had come to see its technical apparatuses as sufficiently reliable to transfer responsibility for malfunction from the instrument to the labourer. Holding the operator responsible for errors was part of the company's strategy for increasing its profitability. The operator who was guilty of what the employer called "unsatisfactory production" was penalized either by being forced to work overtime, or by suffering a decrease of income (BCA, sl. 12, 1897, 70).

Also for the first time, the subjection of the operator to the discipline of the work place included rules related to privacy. Regulations enacted by the company in 1899 made it clear that a breach of privacy was so serious that it would cause "immediate dismissal" (BCA, d 926, 1899, 7). Telephone conversations had become "strictly confidential," and any violation of this rule was to be severely punished (BCA, ls 1899–1910). However, paradoxically, the technology used within the labour process obliged the operator to listen to telephone calls for the sake of the company's business. She was expected to practice *civil* listening, which implied that she ought to attempt to hear only the "sounds" of the callers' voices and to ignore the meaning (BCA, ls 1899–1910). This is a typical case in which the operator was not expected to act as a human being, but to identify herself with an object, a machine. Of course, the fact that the operators were compelled to listen for the company's sake gave them opportunities to listen for their own sake. It seems, however, that the company was so confident in the female qualities of submissiveness and obedience that it did not consider, or did not want to consider, the risk of indiscretion.

In the *Rules and Regulations*, much importance was given to the operator's voice. The company defined a "telephone voice," and encouraged operators to use it. This constituted a first step in the specification of the "telephonic voice," which later figured in the company's hiring policy. Operators received instruction "in the right way of speaking" (BCA, ty 9 (5) 1905, 428). The voice, which had been a relatively neglected aspect of the hiring process, suddenly became an important asset, because the new rules and regulations had reduced personal contact between operators and subscribers to a minimum, and it thus was one of the few remaining human elements that the company could exploit. This was done by disciplining the voice and subjecting it to a definite number and form of sentences used in the labour process. (This particular issue will be discussed at length in chapter 4.)

The elaboration of rules was accompanied by the extension of dividing practices. Two new levels of hierarchy were created: assistant chief operator and supervisor. The training of the operator moved from an informal apprenticeship to planned, formal training. While the assistants helped the chief operator to apply the numerous rules and instructions to the workers, the supervisors were to train new operators. For an applicant to become a "well bred," productive worker, a period varying from three to six weeks was allotted: "The operator's manners had to be groomed; she was made more gentle and refined, quicker and more accurate" (BCA, sf.op 1930). After this probationary period, those who were found unsatisfactory were dismissed. A certain degree of informality in the process of learning allowed the operator to preserve some amount of humanity in the labour process; for instance, her actions were not timed, and her activities were not totally regulated (BCA, d 12016, nd, 18).

Later, the regular operator experienced another type of control through the chief operator and her assistants. Each assistant was assigned a small number of operators and was connected to them mechanically. She could listen to each of them without the operator knowing the exact moment of the interception. This was to ensure that the rules enacted by the company would be respected. An Ontario newspaper described the change that occurred in the operator's labour toward the end of the 1890s.

In the old days in the Bell Exchanges ... the discipline wasn't the rigid thing it is now. [The operator could] snatch a moment to talk to her young man when the subscribers on her circuit [gave] her a rest. [Now], she is not allowed to do it. It is strictly against the rules to do it. If she is caught once she will be reprimanded; if she is caught twice she will be suspended; if she is caught a third time she will be dismissed ... This is what discipline did. It put a table in the centre of the Exchange and connects every operator at the switchboard to the table. Then it got a serious young lady and placed the receiver on her head; and said to her 'watch.' So that young lady is sitting there at this moment, and if she suspects that a private conversation is going on over any line, she will connect herself with the suspected wire and hear every word. The offender will be punished as we have indicated. Again, the operator, after bringing two subscribers together, can hear the conversation if she so desires. That also is strictly against the rules ... The lady at the table, if she suspects a breach of this rule, can detect the offender at once. Punishment follows detection. In addition to these elaborated checks, two ladies – they might be called forewomen – walk up and down the Exchange for the purpose of preventing private conversation between the girls in slack moments ... *The girls then, are automata* ... they looked as cold and passionless as icebergs. *But that is only discipline.* (BCA, d 12016, nd, 18, my emphasis)

Since the task assigned to the operators was subjecting them more and more to the machine, the working period was decreased. This was not an act of charity by the company. Rather, the "cases of fainting and fits" were increasing dramatically. Bell decreased the working period to an average of eight hours per day (BCA, sl.18, 1903, 21), allowed "15 minutes rest" after about two hours' work (BCA, bp 1980, 5), and gave one hour for lunch. If the time for rest or lunch was exceeded, the operator had to work late to make up the time (BCA, sl.12, 1897, 70). The rule was a partial blessing for the operators. Until then, if an operator needed time off, she asked someone to take her place on the switchboard, and paid for this service (BCA, ls 1902). This practice was stopped by management, which maintained that there was abuse in its application and that it resulted in a decrease in operator productivity. Time off was then reintroduced on a standardized basis: *the operator did not have her time off when she felt she needed it, but when the company thought she needed it.* As a manager pointed out, this policy was based on a purely economic motive: "The management realises that it is for better to keep able and experienced operators than to be continually initiating new ones, and that the best obtainable class of girl, well bred and educated, is none too good for this important branch of the public" (BCA, d 12016, 1898, 116).

Standardization of the operator's working conditions extended to her holiday as well. Earlier, operators had had three weeks' vacation, and sick days were deducted from the holiday period. Later, vacation time was decreased to two weeks, and no pay was granted for sick days. This implied a subtle but important change for the operator: she lost the little control she had had over her holidays. Since operators were hired partly for their good health, it is likely that most of them could manage to save more than two weeks' holiday out of three. In addition, given the already low wages, a pay deduction for sick days meant a significant financial loss. With this new rule, then, *the responsibility of time off for illness was switched from a joint liability of the company and the operator to an entirely individual responsibility of the operator.* She lost in all ways in this bargain.

In 1917, the company provided its employees with a pension plan and a sickness-and-disability fund for which the latter did not have to pay. This did not enable the operators to recoup their previous losses, however, as the first plan was valid after at least twenty years of regular service with the company, while the latter was applicable after two years (BCA, ncm 1917c). Since the majority of operators quit after two years of service, either to marry or because they were exhausted, these plans were not advantageous for them. Rather, they favoured male labourers, who, because their work was not as strenuous as that of the operators, worked long years for the company.

In short, the telephone company applied a mechanism of control that forced the operator into individualized responsibility and, at the same time, into standardized working hours and holidays, although collective responsibility and individualized work and holiday periods were more advantageous for her. However, she had no direct means to change this mechanism, which was profitable for the company.

Individualization of working conditions was enhanced from 1902 to 1920. In this last stage, the operators were subjected to a "military discipline" within a highly mechanized and scientifically managed labour process. It was during this period that the telephone developed most rapidly. Previously, the telephone industry had been characterized by a certain "amateurism" due to the relatively small scale of its development. However, mechanization of the labour process accelerated expansion of the telephone system and helped to transform the business into a large capitalist industry. In order to handle an ever-increasing system of communication with an ever-decreasing operator labour force, a "fixed routine" and "strict discipline" were deemed "essential" and were means to be "strictly observed" so as to provide "the maximum of satisfaction to patrons and secure ... for the company the full earning capacity of its plants" (BCA, tg 1 (6) 1909, 5). The success of this program was based on "a system conducted *with military precision in all details*" (BCA, tg 1 (6) 1909, 2, my emphasis), within which small groups of operators were "captained by a supervisor who, in turn, [was] responsible to a chief operator" (BCA, ty 5 (5) 1903, 302).

To maintain this discipline, the system of supervisors, chief operators, and assistants was applied more accurately and rigorously.

The supervisors all wear telephone sets with long cords, which are connected to an overhead instruction circuit, thus enabling any supervisor to communicate instantly with any other supervisor in the same office. A key and cord with a relay attachment enables each supervisor to communicate with any subscribers who may complain to the operator in that division, and also enables the supervisors to deal directly with all supervisors in any of the other offices ... It is thus made possible for subscribers to promptly get into communication with the supervisor and also enables the supervisor in any one office to refer a complaint either in reference to a subscriber's circuit number, or about any operator who is not handling her work in a standard manner. With this system over 85% of operators are supervised as to inter-off irregularities. (BCA, tg 1 (6) 1909, 4–5)

The system was efficient. In the late 1890s, the average time for an operator to put through a call was twenty-five seconds; in the first years of the twentieth century, it took her only eight seconds (BCA, ty 9 (1) 1905, 38). This seventeen-second decrease represented a dramatic increase in operator productivity, considering that, in big exchanges such as those in Montreal and Toronto, each operator handled an average of one thousand calls per day. For the telephone system "to run smoothly," operators needed to work rapidly and with strict attention, and any "vexatious delays" of a few seconds, due either to slow subscribers or to technical problems, were at once passed to the "delayed" operator whose task was specifically to take care of these calls (BCA, tg 1 (6) 1909, 2). Any "unnecessary dealing with subscribers" was "immediately connected with the supervisor." There were special services for various types of calls: a "recording services" for out-of-town calls and directory information, a "trouble office" for subscribers' complaints, a "special operator" for subscribers' changed numbers, and an "information desk" for information other than numbers (BCA, tg 1 (6) 1909, 5). To handle this system "promptly and economically," the company found it necessary to have "uniform rules and regulations," and "detailed reports from which the traffic [could] be studied intelligently" (BCA, tg 1 (6) 1909, 3).

The starting point of this "military system" was the "Instruction School." Bell Telephone Co. opened schools, in Montreal and Toronto,[22] where women acquired the training necessary to become *perfect operators*. As a school instructor pointed out, "The operator must now be made as nearly as possible a paragon of perfection, *a kind of human machine*, the exponent of speed and courtesy; a creature spirited enough to move like chain lightning, and with perfect accuracy; docile enough to deny herself the sweet privilege of the last word" (BCA, sf.op 1930, my emphasis). To attain such perfection, informal training was no longer adequate. In the schools, the operator was forced to trade her human quality of spontaneous adaptability to unexpected circumstances for an irreproachable discipline, a "contemplative attitude independent of consciousness," to borrow Lukacs' terminology.[23]

The first step in the process of admission to Bell Telephone Co.'s private training school was an examination of the applicant "by the teacher as to [her] age, which must be over 17 and under 25, [her] height, which must be over 5 feet, and [her] eyesight, which is tested by reading from an optician's card." She was then "interviewed by the Inspector of Service, and examined as to enunciation which must be clear and distinct, and as to [her] hearing" (BCA, tg 1 (6) 1909, 4). As well, "she must have recommendations and avoid gum-chewing and too many necklaces ... All must have good education" (BCA, nct 1914c). These conditions constituted the basis of uniformity in the accepted group. The aim of the school was to further the similarity in qualities through *drilling*. Much work was needed, though, on this "raw material": "Weeks and months ... of patient teaching and practice [were] needed to round them into form" (BCA, ty 5 (5) 1903, 302). According to L.B. McFarlane, to transform the applicants into willing participants was essential, for "there's nothing so crude as an untrained girl," and for the "public sake ... no crude material is ever put on the switchboard" (BCA, nct 1914b). The training course lasted an average of six weeks and was divided into three classes, during which the pupil was drilled not only in how to handle the technical aspect of her machine, but also in adapting to the psychological stress that she would later experience. Humiliation and degradation were tactics used by the instructors to discipline applicants into becoming mere machines (BCA, ncm 1907e, 6). The language used by telephone-company management, men and women, clearly indicates the objectification of the operator. She was repeatedly described in machine-like terms: the carefully schooled applicant "fit into her groove as nicely as the most delicately constructed piece of astronomical machinery" (BCA, ty 9 (5) 1905, 388), said a manager.

In short, the operator labour force lost more autonomy with intensification of control via more specific rules and more rigid training. The proccesses of objectification and categorization, to use Foucault's words, does not proceed evenly over the years. In the case of the operators, it passed from a stage of loose and elementary apprenticeship to a stage of planned training involving the teaching of telephone etiquette, discipline, and some standard forms of working instituted in order to maximize their production. Rules and instructions merely subjected the operators to a standard working process which facilitated the company's control over them.

The definition of a "good" operator implied the existence of a "bad" operator, which involved the "moral connotation" of "being against the rules." In fact, the moral dimension of the operator's work came to predominate over its strictly technical aspect. This dimension was expressed by L.B. McFarlane in an interview for the *Watchman*. He stated that while the "telephone girl" was "used," she "was not abused." As a producer of telephone calls, the operator was "taught to answer impatience with patience, unamability with kindness and consideration but woe betide the hasty subscriber who forgets that the 'phone is feminine, and delivers himself of a 'big, big D__.'" Moreover, "the manager is a strict disciplinarian, but is perfectly just.[24] The operators must do their utmost duty, but they may not suffer indignity. Whenever any profanity or foul language has come over the 'phone, that subscriber has been promptly cut off if the operator reports him" (BCA, d 12016, 1898, 116).[25]

Thus, the relationship between the operator and the machine was dialectical: the operator was becoming more and more of a machine, and the machine was increasingly considered "feminine" because of the indispensable mediation of the operator, which involved moral values.

The intensified training oriented toward technical skills and moral values turned the operators into individual workers willing to do a good job. They repeatedly used the company's discourse, asserting that Bell was a big family in which each operator had to do her best. An applicant was allowed to leave school only when she could perfectly fit into that "big family" and help to reproduce it. Those who could not come to terms with this attitude were rejected. The operators, however, were conscious of their role as workers and, in spite of the family feeling, responded to certain forms of exploitation in ways that were sometimes unforeseen by the company.

FROM MACHINE TO PERSON

The operator's lack of direct control over the labour process did not prevent her from applying some forms of pressure. While Bell Telephone Co. was disciplining the operators, the latter were starting to show serious signs of resistance. Since control was exercised at the individual and social levels, the operators' resistance manifested itself at both levels too. At the individual level, for instance, personal use of the telephone lines became much more serious: "In the evening, when I wasn't busy on the board, I used to talk to the operator in Port Colborne for something to do. Both she and I would leave the connection up between us while answering our respective customers. We weren't supposed to do this but there were a lot of little things like that that we weren't supposed to do" (BCA, ls 1917).[26] Another form of individual opposition to the company's rules consisted of the exploitation of newly arrived operators. The experienced operators arranged to have the new ones, who were zealous and enthusiastic, do most of their work. The result was a slow-down in service and a decrease in the experienced operators' work load. At the same time, it indicated a certain unofficial hierarchy among the regular operators. The operating of the "nickel telephones" provided another opportunity to resist the company's regulations: "The operators ... seeme[ed] to have a few unwritten rules of their own on the subject." They gave the call free to nice customers, but with a nasty user, "generally a woman," they "simply collect[ed] the nickel and disconnect[ed] the subscriber," who lost both the nickel and the telephonic communication (BCA, ncm 1907e). In short, covert individual practices countered some of the rules imposed by the telephone industry on the operators. They represented a way of applying individual power against a situation which could not directly be changed by the workers.

At the social level, operators organized in an attempt to counter the company's enforcement of control by opposing certain rules and by extracting some benefits from their employer. In 1894, they created the

"Lady Operators Benefit Association," which allowed them to receive a part of their wages after three days of illness (BCA, sl.16, 1901, 9). To counter the standardization of working conditions, they started to organize group resistance. However, this manifested itself only a few times during the period studied, and with mixed results. Operators were not unionized until 1918 and, as we shall see, their union lasted for a very short time. Resistance usually came from small groups of operators in big cities, but these people were never able to present a common front against Bell Telephone Co.

The earliest evidence of a collective complaint from the operators to the company was in April of 1883. The telephone operators in Montreal sent a petition to Bell management asking for better wages and shorter working hours. The demand was denied in a laconic statement asserting that there was "no prospect of the salaries being increased beyond $20 per month" (BCA, d 18919, 1963, 57). The next collective demand was made in 1903, also in the form of a petition, by night operators in Toronto. It was sent to Mr Dunstan, the local manager in that city, and requested higher wages "in view of the additional cost of living" – an increase which, the petition implied, had already been "granted to fellow employees."[27] The operators' petition also demanded shorter working hours, in view of the night work being "more trying to the constitution that the day work." The night operators earned thirty dollars per month and worked from 9 p.m. to 7 a.m. Although the local manager admitted that these were long hours, he recommended to Bell management in Montreal that the night operators' wages and hours "remain as they are for the present" (BCA, d 18919, 1963).

A third request, from a group of four Toronto operators, was made in a letter written to Mr Maw, chief inspector of switchboard equipment in Toronto, in December of 1904. Again, the request was for an increase in wages. This letter was written some time after renovation of the switchboard in Toronto, in 1903, had increased the level of strain and stress on the operators from noise and physical discomfort. The renovation, according to the operators, greatly increased "the complications and difficulties of our switchboard circuits" so that "the work at the present time not only requires constant study and greater technical knowledge but also more strenuous application than at any other period of our connection with the Company." They also complained that their wages were insufficient to meet "the steadily increasing cost of the necessaries of life," and were "beyond the bounds of possibility to make any provision against a rainy day" (BCA, d 18919, 1963, 58). The fact that the Toronto telephone operators went on strike in 1907 for the same demands suggests that they had not been met by that time.

The 1907 Toronto operators' strike was the major protest by telephone operators against their working conditions, and also that which most attracted public attention. This strike has been examined at length from different points of view in several studies (e.g., Heron & Palmer 1977, Sangster 1978), and so I will not go into detail here. However, it is important to situate the strike within the general pattern of telephone operators' organized resistance.

The organization of the 1907 strike has been examined by Sangster (1978) in her article "The 1907 Bell Telephone strike: Organizing women workers." According to Sangster, "the immediate issue precipitating the 1907 strike ... was not inadequate wages: the issue was an increase in hours." In the same breath, however, she asserts that "complaints, such as wage cutbacks, were crucial to the strike," and dismisses Heron and Palmer's claim that the strike was "an outcome of a managerial drive for efficiency" (Heron & Palmer cited in Sangster 1978, 112). All previous requests from the operators in Montreal and Toronto included demands for increased wages. Moreover, some evidence in the *Report of the Royal Commission on a Dispute Respecting Hours of Employment between The Bell Telephone Company of Canada, Ltd. and Operators at Toronto, Ont.* (BCA, d 14420, 1907) (hereafter called *Report on Bell Telephone Operators*) suggests that the issue of working conditions was inseparable from operators' wages and the company's drive for efficiency. A serious analysis of the dispute should examine these issues as different aspects of the same problem. As the commissioners put it in the conclusion of their report, the change Bell made in the operators' schedule, from five hours' to eight hours' work per day with a wage increase of only two to five dollars per month, was "made ... from motives of cost and service pure and simple, and without any real consideration for the health and well being of those whom it was most to affect" (BCA, d 14220, 1907, 95).

The increase in wages did not compensate for the increase in working hours (see table 3), although Bell management denied that the longer hours were based on financial interest. Operators' wages, Bell managers said, were "goods" paid at market price (BCA, d 14220, 1907, 35–6). The commissioners' conclusions suggest that they did not believe the managers, and that they attributed the discrepancy between increased working hours and increased wages to the company's drive for profit *and* efficiency.

An increase in the wage schedule was a necessity, if the company was to maintain its service, for without an increase in wages, operators could not be obtained, and without operators the service could not be kept up. To offset the

Table 3
Proposed Wages and Pay Schedules, 1907

Length of employment	Five-hour schedule		Eight-hour schedule		Seven-hour schedule	
	Amount per month ($)	Amount per hour (¢)	Amount per month ($)	Amount per hour (¢)	Amount per month ($)	Amount per hour (¢)
1 to 6 months	18.00	13.8	20.00	9.6	20.00	10.9
6 to 12 months	20.00	15.4	22.50	10.8	22.50	12.3
12 to 18 months	20.22	15.4	25.00	12.0	25.00	13.5
18 to 24 months	22.50	17.3	25.00	12.0	25.00	13.5
24 to 30 months	22.50	17.3	27.50	13.2	27.50	15.1
30 to 36 months	25.00	19.2	27.50	13.2	27.50	15.1
36 months +	25.00	19.2	30.00	14.4	30.00	16.5

Source: BCA, d 14220, 1907, 33.

increase in cost occasioned by the increase in wages, the hours of service were lengthened, the percentage increase in the hours of employment being made considerably in excess of the percentage increase in the rate of wages. The company sought to bring about the change on the shortest possible notice, and in a manner which affords grounds for believing that it hoped to enforce the new schedule by taking advantage of the necessities of its employees, and the fact that as young women, many of whom were self-supporting, a threat of dismissal would be sufficient to prevent any general or prolonged resistance. (BCA, d 14220, 1907, 95).

Indeed, in an effort to break the operators' resistance, the company compelled them to accept the new schedule in writing or resign at once from its service. The operators' response was to go on strike, since they disagreed not only with the increased working hours but with the supposed increase in wages as well.

Various studies have reported Bell managers' statements in the *Report on Bell Telephone Operators* concerning their agreement on the fact that the operators' wages were "not sufficient ... to really properly pay living expenses in the city of Toronto," at least "not of the class that [Bell] wanted" (BCA, d 14220, 1907, 30). This implied that if the women whom the company hired were not living with their parents, their wages did not enable them to maintain the social status desired by the company. In other words, the company hired women from the lower-middle or upper-working classes who, if they were self-supporting, could maintain or improve their social conditions only insofar as they could gain a surplus of wages by way of overtime hours or a second job.

Table 4
Wages for Five-Hour Schedule and Overtime, 1907

Length of employment	Five-hour schedule plus overtime ($ per month)	Total ($)	Eight-hour schedule ($)	Decrease to operators ($)
1 to 6 months	18.00 + 7.20	25.20	20.00	5.20
6 to 12 months	20.00 + 8.00	28.00	22.50	5.50
12 to 18 months	20.00 + 8.00	28.00	25.00	3.00
18 to 24 months	22.50 + 9.00	31.50	25.00	6.50
24 to 30 months	22.50 + 9.00	31.50	27.50	4.00
30 to 36 months	25.00 + 10.00	35.00	27.50	7.50
36 months +	25.00 + 10.00	35.00	30.00	5.00

Source: BCA, d 14220, 1907, 32.

The five-hour day had allowed the operators to do a lot of overtime, an average of fifty-two hours per month, by which means they could increase their wages by $7.50 to $10.00 per month (BCA, d 14220, 1907, 31). With an eight-hour day, the operators lost this possibility for extra pay, especially since the company was trying to prohibit overtime. Thus, the operators were losing much more than the official figures showed (see table 4). In fact, they were fighting neither for a decrease in working hours in itself, nor for an increase in wages as such, but for decent wages with reasonable working hours which would allow them to be self-supporting. This combination appeared incompatible with the company's never-ending search for more profit and operator efficiency.

The same economic conditions drove the operators close to a strike in 1918, and led them to affiliate with the International Brotherhood of Electrical Workers (IBEW). The operators' attempt to unionize was part of a larger movement in Canadian industry. The adoption, by the Canadian government and some Canadian industries, of the "McAdoo Award"[28] encouraged workers in other industries to request similar increases in wages. Bell Telephone Co. attempted to avoid unrest among its workers by granting an increase in wages that was below the McAdoo Award, but this particularly dissatisfied the Toronto operators. In August of 1918, the four thousand Toronto operators started to organize to demand higher wages. On 10 August 1918, five hundred telephone operators, despite their supervisors' threat of dismissal, met at the Labour Temple and decided to form a local of the IBEW and to demand a weekly wage increase of three dollars. On 13 August 1918, they affiliated with the IBEW; a few days later, Bell Telephone Co. granted them the increase in wages they had requested, but refused to recognize their union. The operators complained about the company's attempts to force the operators to resign from the union and about its threats that it would do everything it could to smash it. These complaints forced the minister of labour to appoint a Board of Conciliation under Industrial Dispute Acts to examine their grievances. Before the Commission wrote its report, Bell and the operators' union signed a first collective agreement which met most of the operators' demands. However, the agreement had no duration clause and, despite some efforts at maintaining and even expanding the union among the operators, membership dwindled, partly, it seems, from a lack of interest on the part of the telephone operators, partly from lack of assistance from the IBEW (BCA, d 18919, 1963, 46–51).

About a year after the agreement was signed, the union was replaced by an association called the Employee Representation Plans, created and sponsored by Bell management. For years, this association was the only forum for worker-management negotiations. The first attempt by Bell management to establish the association, among the supervisors in early 1919, was unsuccessful. A few months later, however, the same association was proposed to the workers, who agreed to affiliate as members. The attractive membership conditions set by Bell did not fail to catch the workers' attention: three months' employment at Bell, no fees, expenses related to its operation paid by the company, monthly meetings followed by meetings of groups of workers with company managers to discuss working conditions and management problems (BCA, d 18919, 1963, 84–5). To foster the "family spirit" necessary to maintain the effectiveness of such an association, Bell managers regularly gave talks at the workers' monthly meetings. As well, the company sponsored sporting associations for the workers, of which C.F. Sise, now president of Bell, became honourary president and several company managers became honourary patrons. Paternalism, the ideology on which Bell Telephone Co.'s wealth rested, was alive and well. The family spirit encouraged by management, coupled with repressive practices applied to the operators, helped to quell any serious resistance movement for a long time.

In this chapter, I have shown that it is necessary to use various theoretical concepts and approaches to understand the complex relationship between the feminization of an occupation and the political-economic development of the entreprise concerned. In the case of telephone development, a political-economic approach emphasizes that the interaction between telephone companies, particularly Bell Telephone Co., and telephone operators was crucial for the development of the telephone business. A feminist approach reveals that, for most of the years of the early development, the "feminine" characteristics assigned to women of that era, exploited to their maximum by management, were largely responsible for expansion of the telephone industry and for its profitability. Managers found the future of their company very problematic with male operators, in whom they could not find the submissive inclinations socially inculcated in the other sex.

Foucault's concept of subjectification helps to clarify the motives behind feminization of the occupation of operator by telephone companies. Indeed, the analysis of the three processes to which the operators were subjected stresses that women were more inclined to subject themselves to the companies' exigencies, albeit within certain limits. As workers, they were willing to comply with their role as defined by the company, but, at the same time, they were conscious of certain aspects of their exploitation and ready to oppose the most unacceptable of them. The response of women operators to the telephone companies' exploitation affected their relationship with the subscribers. This will be discussed in the next chapter.

A.G. Bell's "long goose neck" wooden hand telephone and wooden box telephone of 1877 were the first telephones leased for commercial use in Canada. By moving the instrument from ear to mouth, the user could transmit and receive with either unit. (BCA, d 10716a, 1877–79)

Left: The Blake magneto desk telephone of the 1880s relayed the voice with increased clarity. It was in general use in Canada until about 1900. (BCA, d 2099, no date) *Right:* The White solid-back long-distance magneto telephone of the 1890s was so called because it gave better transmission than the Blake telephone for long-distance calls. (BCA, d 12985, 1891)

With the iron handset, the operator could work with only one hand. (BCA, d 613, 1879)

In 1884, the operator not only had to wear the six-pound Gilliland harness headset for hours, but she also had to work with both hands. (BCA, d 6151, 1927)

Linemen standing on a sixty-five-foot pole with six double crossarms used by Bell Telephone Co. at the turn of the century, St Hyacinthe, Quebec. (BCA, d 13622, 1908)

Linemen installing underground telephone wires, St Clair Ave, Toronto. (BCA, d 3286a, 1913)

In early days, exchanges were often in private residences. The manager is standing behind the operator, supervising her work, Bowmanville, Ontario. (BCA, d 6628, 1899)

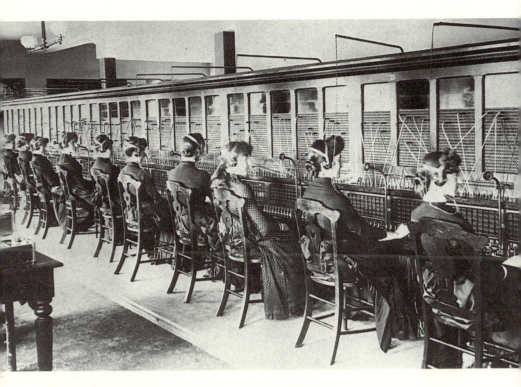

With the first multiple switchboard, working conditions for operators were still relatively acceptable, with reasonable space between each worker, and comfortable chairs. Montreal's Main Exchange, Bell Telephone Co., circa 1890. (BCA, d 3109–1, ca. 1890)

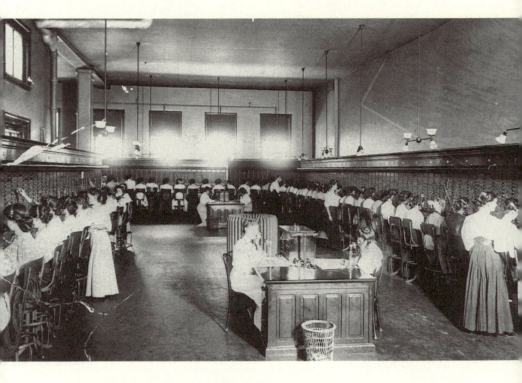

The common-battery switchboard was attended to by "drilled" operators who were exploited to "the limit of human endurance." Toronto's Main Exchange, 1904. (BCA, d 10364a, 1904)

In rural areas, the operator was a "Jill-of-all-trades," lighting the fire, cleaning the office, checking the accounts, repairing the technology, even connecting the calls. Mrs J.P. Johnstone at her desk, with the switchboard and the coal-oil stove in the background, Sundridge, Ontario. (BCA, d 15869, 1911)

Even after the occupation of operator was feminized, male operators were used at the "test desk," behind the switchboard, to check the technology. Montreal's Main Exchange, 1909. (BCA, d 4730, 1909)

A woman subscriber using a public telephone at the end of the period studied.
(BCA, d 26579, 1922)

CHAPTER FOUR

Voicing the
"Pulse of the City"

The relationship between subscribers and operators was the subject of many articles in newspapers, magazines, scientific journals, and other publications during the early period of telephone development. Some described the role the operators played in their gradually changing relationship with subscribers, while others evaluated the participation of operators and subscribers in the use of the telephone. An oft-discussed issue was the voice.

With the advent of the telephone, the voice had become a long-distance agent of sociability. The characteristics of the telephonic voice were replacing those of handwriting, since the former was increasingly perceived as revealing the personality; thus, it became as important to "practise" or "educate" the voice as it was to "form" one's handwriting. Voice teaching was usually recommended for both sexes, but was aimed particularly at women. Special attention was directed, toward the telephone operator. "Your voice and the telephone courtesy are your only *seeing* points in telephone conversation," said a Bell Telephone Co. manager in a lecture to his employees in the 1910s (BCA, d 29547, nd, my emphasis).[1] The fact that the manager used a word related to sight implied that the voice had taken the place of both speech and sight, and therefore had become doubly important. For telephone companies, the voice, as a cultural element, reflected the class of the person to whom it belonged. Courtesy was associated with bourgeois and upper-middle-class manners. A working-class voice could not be courteous if it was not trained to be "lady-like." A rough or high-pitched voice could not belong to someone from the affluent classes. "A test of whether gentility is a thin veneer or the solid substance is that of the telephone voice" (BCA, tg 3 (1) 1911, 5). Identification of a "genteel" voice applied to both operators and subscribers.

"THE WOODSY VOICES OF
A SUMMER DAY"

For operators, the voice itself, and not the worker as an entire being, was the mediating component between subscribers and telephone companies. It was the "agent" representing the class, moral values, and personal characteristics of the operator, although its importance gradually changed over the years and stood in inverse relation to the operator's autonomy. The more mechanized the labour process became, the more impersonal was the contact between subscribers and operators, and the more important the voice.

During the first period of telephone development, when the labour process was still unstructured and the relationship between operators and subscribers was on a very personal level, there was no mention of the voice as a characteristic of the labour force. At that time, the *content* rather than the *form* of the communication was emphasized. Using speech that was spontaneously adaptable to unforeseeable situations was the most important aspect of the operator's work. Courtesy, flexibility, adaptability, and patience were the required qualities, and very little attention was paid to the formal aspect of the relationship between the operator and the subscriber.

Management at telephone companies began to stress the role of the voice toward the mid-1880s, with the elaboration of loosely defined rules to be enforced by the newly appointed chief operator. The chief operator was told that regular operators should master their voices, and that she should "check" them. This matter took on increasing significance in the internal policies of the company, and eventually operators were subjected to extensive voice training. In the earlier period, however, it was suggested only that the chief operator ask her operators to pay some attention to the tone of their voices (BCA, d 24096, 1884) and that she prevent them from conversing with subscribers. The limitation of personal contact between operators and subscribers was associated with increasing standardization of the voice, which immediately became a means of control over that relationship. A "mastered" voice could not and would not let the operators' identity intrude into the means of communication – something that was potentially embarrassing for the companies, as well as time-wasting.

The voice became a component of operators' working conditions through its subjection to the list of official rules and regulations published by Bell Telephone Co. in 1892, in which the tone of voice was clearly specified on two occasions: "Speak in a clear and distinct tone," and "When number is given, repeat it back in a distinct tone, each figure separately" (BCA, d 920, 1892, 2). Henceforth, the operator's tone of voice was limited, as was the content of her communication. Specific and standard phrases were to be used in particular circumstances. When she answered a subscriber's call, for example, she was now to say only "Number, please," while she previously might have used one of a variety of phrases, such as "Whom would you like to call?" She was also required to say "Line busy" instead of "Sorry! The line is busy now," and "They do not answer" instead of "There doesn't seem to be any answer" or "Sorry! They don't answer." The curtness of the recommended phrases, in addition to their formal regularity, not only deprived the operator of personal contact with the subscribers, but of course also increased her rate of production: the shorter the answer, the more calls she put through within a given period of time.

The issue of the voice was not brought to the attention of the long-distance operator in the list of her duties included in the 1892 publication. Long-distance lines were still technically problematic, and conversations rather difficult; thus, the voice was not the most important element of production. Rather, the operator was told to "assist" subscribers in a personal manner – to use "soothing" speech to appease the ire of businessmen, who considered the wait for a long-distance call wasted time and sometimes acted very rudely. The technology partly determined the operator's duty and qualities required of her.

Later, a new list of duties for regular operators drawn up by Bell management made the limitations more formal. Two tones of voice were specified: a "clear tone" for speaking to subscribers, and a "low tone" for conversation between operators (BCA, d 926, 1899, 6–7). The clear, distinct tone was meant to convey to subscribers both efficiency and diffident eagerness; the low, indistinct tone was used to conceal "domestic" contacts between operators from users.

From 1899 on, these steps toward specification of the "telephonic voice" were part of the telephone company's hiring policies: "The girls who work at the switchboard are now chosen largely for their native adaptability and qualifications, among which the possession of a good voice bulks heavily. *They receive instruction, of course, in the right way of speaking*" (BCA, ty 9 (5) 1905, 428, my emphasis).[2] The voice suddenly became an important asset, precisely because the new rules and regulations had reduced to a minimum any personal contact between operators and subscribers: "The only human element that enters into the situation is that intangible thing – the voice" (BCA, tg 2 (9) 1911, 3). This unique human element in the telephone operator's work was to be exploited to the maximum, by disciplining the voice and subjecting it to the use of a definite number and form of sentences. The "telephonic voice" was defined as a "clear voice." As an operator pointed out, "That fascinating huskiness often recognized as a charm would never do in a telephone operator" (BCA, sf.op 1930). Applicants with the proper voice were chosen and put in training schools to "educate" their voice. The operators were drilled to repress any personal feelings in their voices, except those carefully taught by the instructors. Was it to avoid the perception of transience that telephone companies went to such lengths to annihilate variations in expression?

The problem of the operator's voice, from the company's viewpoint, was mainly one of class identification. Most of the operators were from the working class, while most of the subscribers were from the ruling and middle classes. Even when the operators had the right vocal tone, they needed training to change their working-class voices into "well-modulated lady-like voice[s] ... always kept well under control, no matter how trying the circumstances ... [and which] disarm those who are prone to be irritable or fault finding" (BCA, tg 1 (8) 1909, 10). This voice was also called a "gentlewoman's" voice, which was "never to become strident, angry, or high-pitched" – the perceived characteristics of a working-class voice. Hence, the working-class operator was to acquire a bourgeois or upper-middle-class voice, one which successfully "disarms anger, repels insolence, makes the rough place of daily business smooth" (BCA, tg 2 (6) 1910, 11) for the subscribers coming from these classes. To attain this ultimate quality of voice, the operator had to go through a long process in which she was "taught to pronounce each syllabe clearly and accurately, and with the rising inflection that gives a pleasant tone. The little phrase, 'number, please' may be made either very abrupt as [sic] very gracious, depending entirely on the tone" (BCA, sf.op 1930). The operator's voice was thus standardised in form and in content.

Voice and speech played a multi-faceted role in the development of the telephone business. Their limitation helped to increase the operator's production considerably. In addition, controlling the operator's speech in her dealings with subscribers was a way of controlling her moral values. Since the operator was using only standard phrases, there were few occasions when she could employ expressions disapproved of by certain classes of users. Sise emphasized this point in a letter he wrote to the Toronto local manager: "In Toronto ... I was surprised to see with what a very languid and leisurely manner the operators replied to, and made connections for subscribers – a manner which was very graceful, but which reminded me of my Grand-mother playing the Harp; rather than a lot of women paid to do a certain work" (BCA, sle 1895).

The desired degree of efficiency was reached in schools, where "[a] great deal of attention is paid to the method of repeating numbers. The vigorous rolling of r's is discouraged, though the operator likes to do it, it works off her surplus energy. But to roll the 'thr-r-r-ee' until it is exhausted is an unnecessary waste of time. The stereotyped phrases to which her conversation is limited today have been devised as time savers.

In completing a connection, every tenth part of a second counts; she gives her report to the subscriber in the fewest possible words. To say 'thank you' instead of repeating the number saves one more brief second of time" (BCA, sf.op 1930). This instructor did not mention another essential aim of the training: to decrease the number of errors made by operators. Clarity of tone was also a "time-saver" because it tended to eliminate unnecessary repetition. A clearly projected voice and specific phrases comprised an effective means of control over the operator's productivity.

As a newspaperman pointed out, "If ever the rush of girls into the business world was a blessing it was when they took possession of the telephone exchanges ... The girl and the telephone are natural friends"(BCA, ncm 1916a). The perfect operator had a "gentle voice; musical as the woodsy voices of a summer day," "sweetly distinct to the subscribers ... yet ... carefully articulated" (BCA, ty 9 (5) 1905, 388). This voice was supposed to play "a big part in moulding the temper of the time. Irascible, petulant, hurried, the subscriber cannot help but feel the influence of that something which appeals to him as quiet, dignified, soothing, until his temper melts away as the mountain snows before the compelling chinook of the south west" (BCA, ty 9 (5) 1905, 389). This calm voice was also said to possess a magnetism which "men of wealth" and "farmer boys" found equally pleasant. "I met my husband through a blind date, talking to him from the switchboard while on duty," said an operator (BCA, ls 1907–17). The standard romance story was that of a working-class operator who attracted a millionnaire with her voice (BCA, ty 7 (2) 1904, 126–7). Love stories between operators and working-class men were chalked up to the formers' "soft voice," which "ensnare[d] men's hearts." "There is something about the sound of the voice of a girl on the wire that sets a young man into a wooing mood," said the manager of an American independent telephone company (BCA, ty 9 (4) 1905, 328).

Thus, with the telephone the other senses had a secondary role, and the voice, which had hitheto been a relatively neglected human characteristic, took on much greater importance. "Too few of us realize that the voice is all that goes over the wire. It represents the person who is speaking. If the tone is harsh and grating, the smile in your eyes will be lost upon the person at the other end of the line. – Get the smile into your voice" (BCA, ncm 1917a).

THE ART OF TALKING OVER THE TELEPHONE

The importance of the telephonic voice, however, was not stressed only to operators. Advice on the voice to be used over the telephone was also given to subscribers. It was claimed that there was an interplay between the subscriber's voice and the use of the telephone: using the proper voice gave a more satisfactory quality of communication, while regular use of the telephone improved the voice of the subscriber. The telephone voice was considered to be "greatly a question of sex, but not by any means altogether so" (BCA, nct 1913i). There was common agreement on the fact that "it was a difficult art to project one's personality over the telephone" (BCA, tg 3 (1) 1911, 5), but various reasons were proffered for this difficulty. Some argued that it was because "the distinctness of telephone utterance" made the voice reveal all nuances in the user's mood, meaning that it was impossible to hide impatience or anger (BCA, ty 9 (5) 1905, 428). Others, to the contrary, claimed that "the telephone deprives the voice of the speaker of characteristic inflections." As a result, when subscribers did not use the proper voice, "what was tenderly uttered may sound harsh and grating and be misconstrued" (BCA, tg 3 (1) 1911, 5). In any case, both arguments pointed to the necessity of adopting a "telephone voice" when using the technology. Businessmen were told that "a good voice, used intelligently, politely and persuasively, is a commercial asset." Telephone companies advised "those who deal with the public by telephone" to coach "all new employees," women and men, on the proper way of speaking over the phone. Women's and men's voices were said to have different characteristics: "The woman's voice carries better over short-distance connections, the man's over long-distance" (BCA, ty 9 (5) 1904, 428).

Since talking over the phone was an art to be cultivated, and since Americans were recognized being the most frequent users of the telephone, some writers stressed that the telephone had "improved [the] American voice." As a nation composed of "early settlers, intense people, face to face with the most vexatious problems of existence," Americans "spoke out with a certain crude harshness." Sir Morell Mackenzie, "the eminent English specialist," had attributed the cause of this problem to Americans' puritanical minds (BCA, ty 9 (5) 1905, 428). With the telephone, however, "the American voice [was endowed with] a tone of culture and refreshment ... Hypercritical foreigners hereafter will not be able to find fault with the American voice, and call this a nation of nasal-voices shouters, for the telephone is taming down the so-called loud manner of the native American" (BCA, ty 10 (5) 1905, 360).[3] The influence of the telephone was said to be mainly "upon the conversational voice." Regular use of the phone obliged one "to articulate with the greatest distinctions; to be deliberate on what one says and to make ample pause at the end of each sentence" (BCA, ty 8 (1) 1904, 75).

The effect of the telephone on subscribers' voices influenced their relationship with the operators as well. When users clearly articulated the number to be called, the operator could make the connection more promptly, which accelerated production of telephone calls. It was no wonder that telephone companies asked subscribers to speak distinctly when asking for their numbers: under the appearance of an interactive relationship limited to the effects of a technology on cultural practices lay the exigencies of profit.

The voice, then, was an important contributing element to the relationship between operators and subscribers. This relationship, however, was not limited to the quality of voice used on both sides. The operators played several roles as they adjusted their contribution to the demands of customers at the other end of the line. Most of the time, they were helpful, making the telephone more valuable for the subscribers and more profitable for the companies. From time to time, though, the relationship soured.

DEVIL OR HEROINE?

The instantaneous character of the telephone call created a close mutual dependence between operators and subscribers. The operator was in direct contact with the subscriber, and was thus the company's "true representative." Despite her importance to the labour process, the operator did not have a direct influence on its structure, but was herself subjected to it. Her relationship with the subscriber changed dramatically over the periods of development, owing to changes in the production of telephone calls. As production changed, and as this in turn altered the qualities of the operator's working conditions, so did too the nature of telephone use change. That is, the qualities of the operator's labour and the type of the telephone use interacted.

Early telephone communication contained an element of excitement for operators and subscribers, not only because they were experiencing a new technology, but because their relationship was very personal. This relationship was partly due to the fact that, before telephone numbers were instituted in 1883, the operator knew the names of all her subscribers, and the addresses of most of them, and the subscribers knew the names of the operators working in the exchange. "The operators knew their subscribers by name. They took great pride in giving them good service and did not allow adjacent operators to answer any of their favourites," said one operator (BCA, ls 1892). Later, as the labour process was mechanized, it would become unthinkable to "choose your subscribers."

A relationship has a different quality when the parties involved are acquainted by name instead of by number: "Operators and subscribers were in close touch in the early days when names and not numbers were in order," said one operator (BCA, ls 1880–91). Since there were no rules against conversations between subscribers and operators, most subscribers had more or less lengthy chats with the operators, and several of them insisted on talking to "their" operator. In return, the operator of the early periods always talked of "her" subscribers: "We got to know our subscribers very well and did many little turns for them and some of them were exceedingly nice. I had one lady, she was very wealthy, and very exacting, she would not put through a call if I wasn't there to answer her. She would have her maid call first to see if I was there" (BCA, ls 1889–93).

When a regular operator was removed, "many a subscribers [sic] grew indignant ... so accustomed had they become to her knowledge of their requirements" (BCA, ls 1880–91). Actually, this type of personal relationship between operators and subscribers, while not instigated by the company, was certainly encouraged. The telephone systems were at an early stage of development and were very imperfect. The lines were noisy and the service unreliable. Hence, the precarious technical service was compensated for by personalized services provided by the operators. In fact, the unreliability of the systems was in itself a cause of additional personal services. The operator was obliged to verify, each morning, that all of the lines were working. Her "first task ... was to ring up every subscriber in the district with 'Good Morning: this is the daily test. How do you hear me this morning?'" (BCA, sf.op 1930). This routine call constituted a personal contact with the subscribers and meant a daily communication with all of them, even those who did not use their telephone during that day. Moreover, it was predictable that, since the systems were undependable, subscribers made comments on the service, and perhaps took the opportunity to inquire about personal or community matters.

It seemed that the "subscribers were remarkably uncomplaining in view of the service they got" (BCA, sf.op 1930). Although people were reluctant to acquire a telephone because of its technical unreliability, they were attracted by the reliability of the operators' knowledge about the community's activities or by other, more personal services. An operator summarized the situation: "It was largely personal interest between subscriber and operator that kept us going in the pioneer days when we were just feeling our way" (BCA, sf.op 1930). According to her, this personal interest was translated into "a feeling of neighbourliness that is not possible now" which resulted in the development of an active collaboration between subscribers and operators and a certain community of interest. For example, an operator had sent a doctor to a subscriber's house after she received a call from a child saying that "his mother had been badly hurt." Some time later, the doctor called her back to give her more information on the case: "Jones had hit his wife with a frying pan but ... she was all fixed up now" (BCA, qa, ndc, 39).

This relationship did not take place on an "equal footing," but was based on class differentiation. Indeed, since most subscribers were from the bourgeois and upper-middle classes, their relationship with the operator was comparable to that of a businessman with his secretary, or a bourgeois woman with her maid: the subscribers considered the operators to be at their service. As a consequence, according to whether an operator pleased or displeased a subscriber, the latter was nice or rude to her. In newspapers and magazines, the operator was often compared to a "maid", whose best quality was to be patient and submissive. One writer described the operator as a "sweet maid whose skillful fingers weave with plug and jack the magic circle round" (BCA, tg 1 (6) 1909, 11). Still, she had a certain degree of autonomy, as the subscribers were not her direct employers, and a certain amount of power in her role as a gatekeeper to telephonic communications. For instance, if a subscriber was too harsh and rude, she could retaliate by providing slow, poor service, about which the subscriber could do very little. Indeed, some operators admitted that they had some means to resist unpleasant subscribers that were unknown to the public.

When you are exasperated by angry subscribers, you can, for instance, switch three or four couples together ... and then sit and hear them rage at each other. If you're feeling particularly wicked, you can open the listening key and the call circuit at the same time, so the poor subscriber will have a Babel of talk deluging him from the order-wire ... Then, with a self-satisfied grin, you take your fingers off two celluloid buttons – you've pressed down only two, but the more you press down the more fun it is – and the irate subscriber finds himself talking loudly into a windless silence, broken – at last – by a smooth, calm: "What number were you calling, please?" Yes, there are plenty of ways one finds to avenge one's self upon the telephone hogs. You can give them the click of the "busy back," you can ring the bell in their ears, or you can simply – let them wait and rattle their hooks. (BCA, ncm 1907c)

If the retaliation was grievous, subscribers could complain to the company and have the operator sacked. However, most of the time the reprisals were mild, although upsetting to the subscribers, who were powerless in the situation. On the other hand, the telephone company knew that the operator's work was difficult and that some subscribers were too demanding, so it took only serious complaints into consideration. Often, the company publicly defended its operators, sometimes turning the blame on the subscribers themselves in an attempt to "educate" them: "We hope to educate the subscriber up to a realization of the fact that if he will himself take time to give the number to the 'phone girl clearly and slowly there will be averted a large percentage of wrong connections" (BCA, nct 1913d). Throughout development of the telephone system, various instructions were given to subscribers, usually in the form of short statements at the bottom of directory pages, on the best manner to use the instrument. Later, as we shall see, these simple instructions became a systematic "education of the public."

As the operator's skills developed in response to the ever-increasing demand for rapidity in connecting calls, so callers were forced to adopt certains skills in their use of the telephone. In the first days of telephone development, standard patterns for its use did not exist. There were as many different kinds of telephone as there were subscribers, and lines were installed in every imaginable manner. There was no single "proper" way to use the telephone, and each subscriber attempted to get the most out of his or her own system. Telephone users were very limited in number. Tolerance and good nature were generally the rules among early users and operators: "My relations with subscribers were always of a pleasant nature and sometimes, in the evenings, when there was very little to do ... we would talk with them on the line" (BCA, ls 1880–95). As the technology developed, however, so did subscribers' expectations, and various types of subscribers began to be discerned.

Two general classifications emerged in particular: the *ideal* user, who was gentle and nice to the operators, and the *villain*. The differentiation of users occurred early, when the telephone was used more for business than for sociability. Efficiency was expected, especially from business-men, and delays and operator errors often provoked unpleasant reactions: "People say rude things to you, and it hurts" (BCA, ncm 1918g). A large number of the villains, according to one operator, were demand-ing businessmen. She describes her perception of them: "A good many businessmen are so wrapped up in their own concerns that they think the world ought to stop moving when there is something of interest to them at stake. They can't understand how we can dream of attending to anybody else's call when they've given us one. Evidently they are kings in their own circle, and they are used to having everybody in sight jump around at their beck and call" (BCA, ncm 1918c).

The appearance of the villain subscriber raised the question of moral behaviour for users. Some of them used "profane and abusive language" with the operators to get their way. In such cases, the latter were instructed "to cut the telephone out" (BCA, sb 80910b, 3144–1, 1881a, 139). Profanity became such a serious matter in the consumption pro-cess that, in 1915, a law was passed to provide for levying fines against offenders. "Swearing" on the telephone was penalized by a "$25 fine, or 30 days in jail." This legislation "emanated from the Ontario Railway and Municipal Board, and probably came from complaints made by smaller telephone companies through the country," said a Bell manager. Bell had already started to react against such practices "by removing a profane man's telephone" (BCA, nct 1915a). The law was not limited to calls to operators, but extended to other subscribers as well: "In the country many of the telephones are party lines and profanity over these lines is most objectionable." The use of "villainous language" over the telephone lines was considered "a cowardly act," because telephone etiquette required the identification of the caller at the first contact (BCA, nct 1916c).

Not all cases of "misuse" of the telephone companies' lines were so serious, though; most of them were simply tedious. For instance, an operator lamented that certain types of women were tiresome subscribers because they were either captious – "ask[ing] more questions to the square inch ... that a man could think of in a mile" – or nebulous – "Some women ... assume that the operator at the other end is an authority on all things telephonic" (BCA, ncm 1918c). These two types of callers were resented by the operator, as they monopolized and congested the lines. Although the operators enjoyed chatting with subscribers during quiet periods, they were obliged to be fast and accurate during busy periods. At these times, another type of villain was also apprehended: the inarticulate user who "simply have lost the art of distinct speech," and whose language degenerated into "something no one could catch." According to a Bell manager, this caused many errors; although the telephone industry dealt with most of its problems by training the operators "in the most perfect manner," delays caused by subscribers were strongly resented, by both the company and the operators, as they considerably slowed the production of telephone calls.

It was "a difficult matter, however, to reach the hundreds of thousands of telephone users ... and to show them the necessity of speaking distinctly, and always directly into the telephone mouthpiece. If the public would learn to co-operate with operators they would certainly find a great improvement" in service (BCA, ncm 1919a). Here, quality of service was related to subscribers' participation. Bad subscribers were partly responsible for filling the operator's job with "a good deal of a nervous strain," until the labour process was organized so that the operator could switch the most difficult cases to someone else. Bell Telephone Co. management attributed the strain to which operators were subject to the subscribers rather than to the pressures of rapid production. If the operator had been allotted the time necessary to cope with difficult subscribers, she would not have experienced the terrible stress of being reprimanded for not coping with these situations. Notwithstanding all the troubles occasioned by unruly users, according to

the operators, ideal subscribers were a large majority. They were courteous on the telephone, affable and talkative with the operators, and
some of them were generous, sending gifts at Christmas or on other
occasions.

Subscribers, in turn, spoke of two categories of operators: the *heroine*
and the *devil*. The heroic operator prevailed during the early periods of
development of the telephone, when an operator's zealous behaviour in
the context of a personal relationship with the subscribers would literally save lives.

During these early days, telephone service took the form of a "public
service": the operator acted as a community animator, informant, con-
ciliator, and so on: "The 'Hello Girl' has been nurse to the ill, mother
confessor to the worried, and policeman, fireman and detective in emergencies" (BCA, sf.op 1941). Her day's work included not only "put[ting]
the calls through," but also taking messages – "I'll be at my sister's for
the day if there are any calls for me" – and answering queries – "Yes,
Mary said this morning that Bill is better" (BCA, d 29909, 1967, 41–
1). Operators acted as a sort of local information centre, giving out
election returns, dates and results of sporting events, and so on (BCA,
ls 1880–91); as confidants – "People ... imagine they had to tell us
their trouble and their joys. We were the recipients of business and love
affairs, family troubles, etc." (BCA, ls 1896); as baby sitters – "It was
about 11:45 p.m. when a little child came on the line crying [because
she could not find her parents] ... I said ... 'you go to bed, and I will
watch over you. Leave your telephone open.' So she did leave her telephone open, and everything was quiet I should say for another 40 minutes or so then I heard her parents arriving and the phone was closed"
(BCA, ls 1900); and, sometimes, as arbiters – Party lines with as many
as sixteen subscribers "produced many a squabble" for which the
"operator was often referee" (BCA, ls 1899–1910). The most popular
service among all, however, was "giving out the time."

The eternal question of "time, please?" grew to such an extent that it interfered
with *legitimate* business. It was so useful a little privilege and subscribers
availed themselves of it so freely I often wondered it didn't interfere with
watchmakers' profits. Businessmen found it simpler to ask the operator the
time than to glance at their office clock. Commuters who dared not trust
themselves to timepieces depended on the telephone to make their trains.
Meals were ordered and served by telephone time; husbands were chided for
their deviations from it; all enterprise, public and private, in the city ... seemed
to regulate its schedule by the infaillible telephone clock. (BCA, sf.op 1930, my
emphasis)

The number of requests for the time increased to such an extent that in the 1920s the telephone company decided to install a special service solely for that purpose (BCA, nca 1923). In addition, operators were asked for other types of information, such as the state of the roads (especially in winter), the location of a fire or a flood, and where to find help in emergencies.

It was particularly in the latter situations that the operator won her reputation as a heroine. She was so knowledgeable in terms of subscribers' names, addresses, and idiosyncracies, and so willing to help, that in an emergency she could find people on the basis of very limited information. She saved lives by knowing where to reach the doctor, or how to reach the fire department. Such achievements were profusely reported in the press, especially when an operator was directly involved in a life-saving process – when, for example, an operator gave a woman first-aid instructions on the telephone to save a poisoned child; or when another operator assisted in the capture of a thief by acting as a intermediary between policemen and witnesses on the lines; or when a third lent a hand to rescue a lost airplane crew; or when a fourth helped to save several lives in a city in flames by staying at the switchboard as long as she could to warn as many people as possible. As one operator pointed out, "She [the operator] shares every crisis that occurs in the region where she works. The giant exchange is the pulse of the city; the first throb of disaster or joy is recorded here. Storm, fire, war, death, every emergency from the birth of a child to a stock panic announces itself by telephone, through the operator" (BCA, sf.op 1930).

Thus, the operator did not lack opportunities to show her ability to handle pressure in emergency situations. Sometimes, the heroine's enthusiasm and sense of duty went too far and she lost her own life (BCA, sf.op 1941). In the Beacon Arms Hotel in Ottawa, an operator on duty at the hotel's private exchange "refused to leave her switchboard until she had warned all the guests of a fire in the hotel," but she herself was burned to death (BCA, qa 1973). A more famous case occurred in Folsom (United States) in 1908, when an operator working for American Bell Telephone refused to leave her switchboard until she had warned all of the people living in the small town about a flood, in which she was herself drowned. This woman was the subject of a poem in the British magazine *Punch*.[4]

The heroic operator followed the rules and was devoted to the company and her community in normal times, but saw clearly when to act against these same rules in rare cases of emergency. Company managers always joined the rest of the population in praising the heroic operator's action, even when it constituted a breach of the company's rules. Bell

Telephone Co. established an award, called the Vail Award after the president of American Bell Telephone Co., accompanied by a prize of one thousand dollars, for these kinds of actions (BCA, sf.op 1941). The operator's most spontaneous actions were thereby subjected to the telephone company's norms.

The diabolical operator, in contrast to the heroine, obstructed consumption of telephone service. Subscribers considered her behaviour not "ladylike." She was impatient and rude with users and had a "character of her own," which meant that she was not submissive and well disciplined. A subscriber described the "devilish" behaviour: "It is only a few weeks since we had an occasion of reporting this branch of your business ... But we must again draw your attention to the fact that although we are getting a better service, we have to take a very great deal of impertinence from your operator in this town which we will not tolerate al all" (BCA, d 29547–1, 1904).

Instead of being helpful, the diabolical operator, intentionally or not, created delays in telephone use by occasionally resisting certain rules and instructions. It is important to stress that the operator's resistance was occasional. Had it been concerted, she would have been sacked by the telephone company. However, she was usually a perfect operator, who sometimes became what customers called a "perfect little pest."

At first, such marginal behaviour was reported by subscribers directly to the company, usually by letter. These were answered personally in letters from Bell managers, usually hand written, explaining the cause of bad service and apologizing if necessary (BCA, d 29144–38, 1888). Later, when the work of the operator became more organized, subscribers' relationship with the company became more impersonal; several complaints reached Bell through local newspapers, via either letters to the editor or articles (BCA, ncm 1908e). In any case, complaints about telephone service remained a constant concern for telephone companies, especially Bell.

POLICING THE DEVIL

Over the years, subscribers' complaints to Bell Telephone Co. regarding operators' misbehaviour, and malfunctioning of the telephone system in general, were increasingly numerous. Bell finally established a "complaints" policy, which came to be considered an essential component of the company's development. For management, the way complaints were handled could made the difference between keeping and losing not only the customers who were complaining, but, perhaps, potential subscribers among friends and relatives. Thus, the company paid close attention to the "manner of handling complaints." Subscribers' grievances were divided into three categories: those that were justified because the company was at fault, those that were caused by misunderstandings, and those that were unjustified. Regardless of category, the general rule was that all complaints were to be "treated with courtesy," even those coming from "unreasonable and uncourteous subscribers" (BCA, d 8353, 1918, 1).

To protect itself, Bell ordered employees to respond to complaints, even the written ones, verbally, "unless they [were] of a sufficient serious matter to require a written reply," in which case "the letter [should] not contain statements prejudicial to the company's interests." For verbal complaints, employees were also advised to avoid "giv[ing] the subscribers the impression that the complaint [would] not receive attention," and to avoid "attempts to give the complainant an explanation ... [which] might embarrass the company in making final reply." The rule was to use a statement that was precise and vague at the same time, such as "The matter will be attended to immediately and you will be advised." Written complaints were to be "read carefully" and promptly answered, to advise the subscriber "that his complaint [was] receiving attention." The company specified the style as well as the stock phrases to use in dealing with complaints involving the traffic, plant, commercial, and other departments. When Bell was at fault, and apologies were demanded, the matter was to be referred to the division manager. In any case, "every complaint, written or verbal, [was to] receive the fullest consideration." By so doing, the company prevented grievances from reaching "the Municipal Councils, Boards of Trade, and other Public bodies" (BCA, d 8353, 1917, 1–3).

Such attention was prompted by managers who thought that the complaints were good for the company. They were seen as "angels in disguise," and played the role of "supervisor," "monitor of the business," and "thermometer ... indicating the temperature and pulse of [the subscribers'] patronage" (BCA, tg 2 (10) 1911, 12). In fact, complaints provided an additional means of control over employees, especially the operators, who were in instant and direct contact with users. A local manager summarized the company's policy: "A loyal dissatisfied subscriber is our best advertising medium. A good paying subscriber is our financial medium. The link between the two is good service, in reality, the only thing we have to sell." However, "to turn a dissatisfied patron into a well-satisfied subscriber," the local manager had to be "a student of human nature" and be in "perfect control of himself" (BCA, tg 1 (13) 1910, 1).

The extensive literature produced by Bell Telephone Co. on the subject of "handling complaints" and "saving subscribers" demonstrates the importance given the matter by management. The reason was that since there were technical limitations to the size of its market, the company had to keep the small number of subscribers it had. Further, its product was a "service" provided mainly to the business classes. The company was eager to please this clientele not only because it was most profitable, but also because it was a matter of businessmen understanding other businessmen. Complaints from rural areas, or from the working classes, about the lack of service did not receive much consideration. Complaints from women were also taken lightly. Several of them were published in a joking tone in telephone journals.

The most frequent complaints involving the operator's work were about wrong numbers, long delays, and misdirected calls occasioned by operator carelessness: "The telephone has become a martyr ... a reform is necessary" (BCA, ncm 1908b). Other complaints regarding operator behaviour concerned "readiness to report 'line busy'" (BCA, ncm 1907c) and "a stolen (therefore fearfully and deliciously sweet) conversation to her companion at the switchboard" (BCA, d 12016, nd, 18). Often, the language used in articles made the operators look even more devilish – "sweetness of stolen conversation," "*one* of her gentlemen friends," and so on (BCA, nct 1914c, my emphasis). Sometimes, however, a diabolical operator's behaviour resulted in dramatic situations. An operator in Toronto, for instance, refused to give the fire department to a subscriber who was too excited to remember the number. "Three times he made the excited request, but on each occasion he was told that the operator could not give him the fire call unless he furnished the number" (BCA, nct 1914a). It was to be expected that such behaviour would cause ire among users.

In fact, it seems that the operators occasioned only extreme emotions, as they were decried with as much passion as they were praised. Letters to the editor in various newspapers were quite sarcastic. For example, one subscriber described an operator as "really deserv[ing] public recognition" because she "scarcely ever gets the right numbers. Sometimes she makes two mistakes, sometimes three, and this morning she made six mistakes before she got the right number. It was a labor of her [sic]. She was good natured about it, but by the everlasting hills I swear she deserves some distinction for stupidity" (BCA, ncm 1919d). Such accusations were often answered by the operators themselves, in an attempt to explain some hidden facets of their work: "I am sure if the case were looked-into they would find that he was to blame for when you give your number the operator repeats it, and if she repeats it wrong you are supposed to correct her. I am sure no girl repeats the wrong number six times" (BCA, ncm 1919f).

Although these various behaviours opposing the rules were not applied regularly, they gave operators some control over consumption of the telephone service. Even the "perfect operator [is] human, and apt to get hungry. After a while your nerves get fraggled ... You begin by being absolutely impassive and impersonal, you work like a machine, but by 12 o'clock, after a bad forenoon of arrogance and egoism ... you grow actually murderous. You take things pretty seriously" (BCA, ncm 1907c). These behaviours expressed the antagonism felt by some operators due to the class differentiation entailed in their close relationship with subscribers. Working-class operators used the means at their disposal to protest against the behaviour of some ruling-class users who considered them their servants.

A much-practised violation of the perfect operator's code was "listening on the line." Despite all company rules and regulations, eavesdropping seemed to be one of the favourite pastimes during lulls: "I had plenty of time for listening, and it was so exciting sometimes that I hated to stop long enough to answer another call" (BCA, ncm 1907c). Most operators, however, denied listening on the telephone lines. With mechanization of the labour process, the company's rules stressed the extreme importance of privacy of telephone calls. Already, in the Victoria Act (1880), it had been clearly stated that "any person who shall wilfully or maliciously ... intercept any message transmitted thereon, shall be guilty of a misdemeanor" (BCA, sf.il 1880a). On the other hand, the criminal code specified that "[e]veryone is guilty of an indictable offence and liable to two years' imprisonment who wilfully ... prevents or obstructs the sending, conveyance or delivery of any communication by any such ... telephone ... Everyone who wilfully, by any overt act, attempts to commit any such offence is guilty of an offence and liable, on summary conviction, to a penalty not exceeding $50, or to 3 months' imprisonment with or without hard labour" (BCA, sf.il 1880b).

These were general regulations applying to all communication workers – telegraph messengers and telephone operators. However, in 1917, the federal Telephone Act prescribed specific penalties of "$100 or 3 months' imprisonment or both" for offending operators (BCA, ncm 1918c). Similar acts were passed in Ontario in 1917, and in Quebec in 1918. In Quebec, the bill was first brought in after an incident in Montreal revealed the practice of eavesdropping by Bell Telephone Co.'s employees. "In Montreal, a citizen was telephoning or being telephoned to, by a lady. The conversation was interrupted by an official of the Bell Telephone Company, who gave the citizen a severe talking to, and an operator broke in to aid the official in his attack" (BCA, ncm 1918b). Not long after, "a bill was introduced in the Legislature Assembly by Denis Lansey of Montreal" (BCA, ncm 1918a).

With such penalties awaiting them, it was no surprise that when eavesdropping operators were asked publicly if they performed such infringements, they denied it: "I never listen on the lines; it isn't good manners," said a laughing operator to a journalist (BCA, ty 7 (6) 1904, 418). Yet when these same operators talked about their subscribers, they revealed things which they could only have learned "on the lines": the butcher's trouble with his customers and how "he gets more roasts than any one I know"; a woman who "is making all that fuss just because that butcher sent some meat up late"; a subscriber's love affairs (BCA, ty 7 (6) 1904, 418). However, there is no evidence that operators, or telephone companies for that matter, were at any time during the period studied penalized for these infractions. It seems that the imperfection of the technology and the necessity "to check the lines" from time to time constituted sufficient protection against any suit. As one manager pointed out, "If she [the operator] listened, it was only to be of more service to you" (BCA, nct 1952). This technology, which was supposed to allow privacy on a public system, had a serious flaw in that human mediators were technically necessary to production of telephone calls.

In the communication industry, the exigencies of capital sustaining ever-increasing development of more sophisticated technology affects not only the process of production of a two-way means of communication, but the process of consumption as well. The transformation of the operator's work from a personalized to mechanized labour affected use of the telephone. Efficiency, defined in terms of the speed and privacy essential to financial transactions, gradually suppressed the desire expressed by some subscribers to preserve a more personal relationship with operators. Capital accumulation was the motivation behind the mechanization and subjection of both the operator labour force and telephone users.

The discussion also points to the notion of resistance involved in Foucault's process of subjectification. The operators did not accept these economic exigencies without protest. They used the little room they found to "personalize," in some way, their production of telephone calls, which affected their relationship with the subscribers. For years, women operators were the main asset of the telephone business, for they used their particular skills to give a specific use value to the telephone. The change in the operator's role at Bell, from "voicing the pulse of the city" to "working like a machine," was related to a shift in the political-economic conditions governing development of the telephone. Bell's shift, from a small, entrepreneurial business to a large, private monopoly, emphasized the need for increased production and led to depersonalization of the operator's work, which had the advantage of diminishing the class differences between operators and subscribers: a piece of machinery does not have class characteristics. The period of transition from a pre-telephone to a telephone society was very much influenced by the operators's participation in the development of telephone systems. Operators helped subscribers with their struggles in the stages of apprenticeship necessary for the acquisition of new communication practices. This will be discussed in chapters 5 and 6.

CHAPTER FIVE

Bridging the Gap between the Victorian and Modern Eras

Although, for some operators, the development of the telephone was a "wonderful adventure in the development of civilization," it had a different connotation for subscribers. The ahistorical contemporary mind perceives the telephone as a "universal," inconspicuous, and inexpensive technology which has always been accessible to the entire community. However, as Marvin points out, the telephone of the nineteenth century "was not a democratic medium" (1988, 153). On the one hand, its price prevented a large portion of the population from having regular access to it. On the other hand, even those who could afford it did not readily adopt the new technology and adapt to its particular characteristics.

The social distribution of the telephone greatly influenced the emergence of communicational forms, while some types of uses, especially women's use, in turn influenced that distribution. The apprenticeship of late-Victorian women in the use of the telephone was class related. At the production level, the development of the telephone involved mainly working-class women; at the consumption level, its social distribution almost completely excluded that class. The telephone, as developed by private capital, had not been planned to accommodate working-class women, although some of them might have benefitted from its use. Women working in home industry, for example, had little time and opportunity to leave their homes and contact people outside, even in emergencies. On the other hand, population density in working-class areas was very high, and some women were able to get the latest news from their neighbours. Women from wealthy classes had different opportunities to maintain contact with the outside world. Most of them had a maid to do errands and to bring home news from the community, or to care for the house when they decided to go out.

Although Bell Telephone Co. has consistently claimed that the telephone system developed "naturally," the system was planned to make

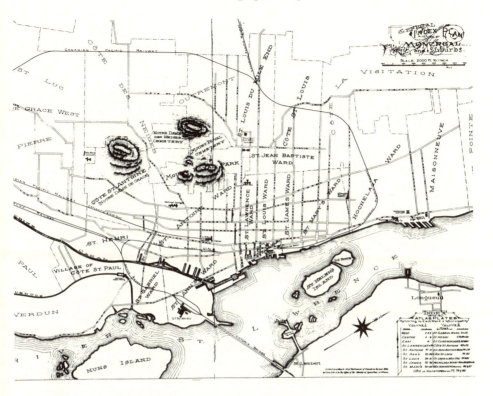

Figure 2: Montreal Wards, 1881–91. (MMA, 1881–91, Goad, nos. 1–2, plate 50)

money. The social distribution of the telephone was limited to particular classes for economic and political reasons. In a "natural" expansion, its distribution would have started from the first exchange, spread on the basis of population density, and served all of the people in the areas crossed by its lines. In fact, as the case of Montreal shows, the system spread along particular social and class axes.[1]

CLASS AND GEOGRAPHY
IN THE DISTRIBUTION
OF THE TELEPHONE

According to the Canada Census of 1881–1891, Montreal was divided into seven wards: East, Centre, and West (ECW) formed one ward, and the others were St Lawrence, St Antoine, St Ann's, St Louis, St James, and St Mary's (see figure 2).

About 70 per cent of the Montreal's inhabitants belonged to the working classes.[2] This majority lived in densely populated wards, in the eastern part of the city – St Louis, with 117 people per acre; St James, with 96; St Lawrence, with 67; St Mary's, with 63. However, some parts of these wards were inhabited by the dominant and middle classes, so that the real density in working-class areas could go as high as three hundred people per acre, and averaged two hundred (Copp 1974, 25), while St Antoine, occupied almost entirely by the wealthy classes, had a density of forty-seven people per acre. St Ann's, peopled mostly by workers, had also a low density (thirty-five) as most of its area was occupied by the Grand Trunk Railway and several types of industry. Finally, the ECW ward comprised the financial centre and Montreal's international harbour. I will examine development of the telephone system on the basis of these geographical divisions, and so it is essential to give a short sociological profile of each ward.

The south part of St Antoine ward, adjacent to St Ann's, contained several large clothing and shoe factories. The eastern part, like the eastern part of St Ann's, was occupied by tradesmen and dealers. The remainder of both wards was inhabited mainly by English bourgeois families in the northern part of St Antoine ward, and by French working-class families, in the southern part. St Ann's was mostly Anglophone, occupied by Irish people who had immigrated during the first half of the nineteenth century. St Mary's, in the eastern part of the city, contained a number of shoe factories. This ward was almost entirely inhabited by French-speaking workers. The industries in St James were mainly foundries and wood-mills, and its population was mostly Francophone too. St Louis comprised the Bonsecours Market, the offices of the Grand Trunk Railway, and some breweries and warehouses. It was largely Francophone, although Anglophones occupied a small section in the north. St Lawrence contained a concentration of clothing and other types of dry-goods stores. This ward divided Montreal's Anglophone west from the Francophone east, and was dominated by small storekeepers; its population was about 75 per cent English speaking. The class distribution in the last three wards – St James, St Louis, and St Lawrence – was oriented on a north-south axis. The working classes were concentrated in the areas south of Sherbrooke Street – especially south of St Catherine Street – "below the hill,"[3] while the ruling classes lived "up the hill," north of Sherbrooke Street. In Montreal at the end of the nineteenth century, French-Canadians formed about 62 per cent of the population, and English-Canadians 35 per cent. Other minority groups were almost nonexistent (DeBonville 1975, 31–2, 215–17).

This profile of Montreal shows that big industry was flourishing in the west of the city and that less prosperous businesses were established in the central and east areas. The English bourgeoisie was concentrated in the northern part of St Antoine, while a small portion of the French middle class lived in the northern part of St Louis. Other groups from the dominant and middle classes were living in the northern parts of these wards. The relations between the classes were very well defined. For example, a gate separated the English bourgeois area in the west from the French-Canadian working-class areas to the east. When labourers from the east had to go west, they had to pay a fee to pass through the gate (MMA 1942).

It is useful, at this point, to mention the living conditions of the working classes. Several socio-economic factors influenced working-class life: income, periodicity of employment, structure of consumption, and environmental conditions. Toward the end of the nineteenth century, Montreal was well on its way to becoming an industrial city. In 1891, there were 35,746 made and female labourers working in 1,604 factories.[4] The sexual division of labour was based on the type of establishment: 82 per cent of the female labour force was employed in eleven types of industry, with most of that percentage concentrated in three types of poorly paid cottage industry – curtain tailoring, shirt making, and shoemaking. Women's and children's wages, although ridiculously low, made an essential contribution to the survival of the working-class family. In 1889, men's weekly wages varied from $4.80 to $18.00, women's from $1.50 to $7.00, and children's from $1.50 to $5.00 (DeBonville 1975, 33). Because of seasonal unemployment, however, a skilled worker could rarely earn more than $300 per annum (DeBonville 1975, 91), and unskilled workers earned much less. Average working-class income during the last quarter of the nineteenth century was quite constant; nominal wages stagnated between 1892 and 1899 (DeBonville 1975, 86). Not only were wages low, but their payment was irregular. Monthly wages, the usual means of payment at that time, favoured super-exploitation. Often, employers delayed payment for months, which put even more economic pressure on working-class households. "When a labourer was working several weeks without wages," says DeBonville, "he [was] linked to his employer until the latter pa[id] him back" (1975, 83).

In 1899, the average annual wage for a working-class family was between $170 and $558, up to 65 per cent of which was needed just to purchase basic food stuffs (DeBonville, 1975, 89–91). For a family earning forty-five cents per day, butter, at thirty-five cents a pound, eggs, at twenty cents a dozen, and flour, at $2.25 per hundred pounds, were almost out of reach. A telephone cost fifty dollars per annum, payable in advance. For a working-class household to have one was completely out of the question, although it was technically possible, since lines passed through the working-class neighbourhoods.

Working-class families spent as much as 25 per cent of their wages on rent (DeBonville 1975, 105). About 10 per cent was needed for even mediocre clothing (DeBonville 1975, 135). There was not much left over for such items as furniture, tobacco, medicine, personal-hygiene articles, or entertainment. "It is enough to read these details," said Jean-Baptiste Gagnepetit, a journalist at *La Presse*, "to see that expenses were reduced to the most basic necessities of life."[5] A large portion of working-class families could not meet municipal-tax payments. In 1886, four thousand families (out of twenty-six thousand) were deprived of running water because of their inability to pay the water tax. In 1890, sixteen thousand warrants of possession were made out for non-payment of municipal taxes. Families thus importuned had to choose between finding another place to live and borrowing money. In the latter case, the only form of credit available to the worker was through pawnbrokers, whose interest charges were usurious (DeBonville 1975, 106–8). "In the whole country, and especially in the city of Montreal, cases occurred where the interest rate per annum was as high as 3,000%. There was, some days ago, in Montreal, a remarkable case where a man who has borrowed $150 was sued and sentenced to pay, in interest only, the sum of $500 on the capital."[6]

Living conditions under such economic constraints were miserable. Working-class families had to pay high rents – five to six dollars per month (DeBonville 1975, 106) – for shabby, dirty, badly built, and unsanitary houses (Copp 1974, 25–6; DeBonville 1975, 106). Population density in some low-income areas was very high: in St Ann's each person had 0.72 rooms, in St Mary's 0.88; in St Antoine, each person had more than two rooms – 53 per cent of the houses had more than eleven rooms! In the working-class wards, up to eighteen persons were sometimes crammed into a six-room house (DeBonville 1975, 116–21).

Hygienic installations were another indicator of wealth. High-income groups could afford indoor toilets, while most working-class families had to be content with pit privies (DeBonville 1975, 107–9). The number of pit privies increased as one walked down the hill toward the southern areas, where French-Canadian and Irish families lived. Although, in 1891, "St Ann's had the nauseating record of the largest number of pit privies per inhabitant" in Montreal – 15.8 persons per pit – pits were thought to be "a barbarous and anachronistic reality" in St Antoine (De Bonville 1975, 124). "The municipality was largely responsible for deterioration of the sanitary services," says DeBonville, since the sewage system in Montreal was "decrepit" and "inefficient" (1975, 126). And the misery of the working classes did not end there. As a general rule, working-class families lived near industrial areas that were polluted by noise, smoke, and fumes. In addition, people living in the eastern part of the city suffered not only from pollution in their own area, but from that from the western sector as well.[7]

In short, the poor living conditions of working-class groups were typified by small and cramped living spaces, lack of sanitary facilities, an inefficient municipal sewage system, and industrial pollution. While employers were responsible for a part of that misery, since workers' wages could not support a decent living standard, governments – federal, provincial, and municipal – could also be held to account since they did not enact regulations and laws that could have improved the lives of the these people. A government that allowed a majority of its population to live in such impaired conditions without providing sanitary and health facilities to the most deprived groups was not likely to supply them with a "luxury" commodity such as the telephone.

The telephone system was developed for and on the basis of a business clientele. The obvious utility of the *residential* telephone was considered to be the link if afforded between the businessman's office and his home. Conversation on telephone lines for any purpose other than business was seen as a waste of time and money for the telephone company and its subscribers (BCA, d 30114, 1965). The telephone interested this class of subscribers because it allowed them to save time and labour. It is not surprising, then, that in all Canadian and American cities the first exchanges to be opened were situated in the business districts.

In Montreal, in 1880, what was to become the Main exchange was located in the heart of the financial district. This exchange served about ten square miles, covering the whole financial area as well as the industrial sector south of St Antoine, and a part of the bourgeois neighbourhood around Dominion Square and St Peter's Cathedral. There were 246 subscribers: 196 businesses and 58 residences.[8] The number of residences is misleading, however. A number of small businessmen, such as grocers, tobacconists, druggists, and so on, had their living quarters above their shops, so that the telephone was accessible to members of their households at least in emergencies. There were no telephone numbers in the first directory to serve this exchange. Subscribers identified the connection they wished to have made, by name and address. The first telephone numbers appeared in the Montreal directory of 1 January 1883 (MMA 1883).

In 1887, before a second exchange was opened, there were 2,082 subscribers in Montreal (BCA, slo.2, 1887). The Uptown office was also opened in 1887. It was situated in northern St Antoine and served the ruling classes in the northern part of the ward, and the big industries in the southern part. A third exchange, East, was opened in 1888 a few streets east of St Lawrence Street, the central axis of Montreal, just north of the most populous working-class area in St James and midway between the foundries and wood-mills to the south and the dominant and middle-class neighbourhoods to the north. In 1890, a fourth exchange, South, was opened in the northwestern part of St Ann's to serve the factories in the area. A fifth exchange, Westmount, was opened in 1898 in the wealthy Anglophone area of St Antoine. In 1899, after South exchange closed, because of a lack of promising customers (BCA, d 3521, 1899; sb 80910b, 3144–1, 1892), the four remaining exchanges served a total of 3,622 subscribers in Montreal. Residential telephones numbered 941, a little less than 26 per cent of the total. Finally, in 1903, still with four exchanges in operation, there were 3,403 residential telephones in Montreal – less than one per one hundred people (BCA, d 2935, 1903). Similarly, Toronto had 3,066 residential telephones – about 1.4 per cent of the total population of that city.

Since the telephone lines connecting the four exchanges had to cross low-income sectors in order to reach either the industrial or the ruling-class areas (see figure 3), a "naturally" expanded telephone system would have included the former areas. An inexpensive party line erected in these areas for working-class families could have lessened the burden of their misery. There were several uses for a telephone in a working-class household. Because women had to work to supplement men's wages, children were often left at home alone with no or little supervision. A telephone could have been used for the mother to check on the children from time to time. Women working at home as part of the cottage industry could have used the telephone to communicate with their employers, or to do their errands without having to leave their work. Since unemployment was continually striking these workers, the telephone would have been of use in job-hunting. In fact, a company manager told a journalist that it would have been cheaper to provide working-class families with inexpensive telephones than to run lines through the areas and install only public telephones that cost five cents per call (BCA, d 12016, 1879–1906).

Public telephones were very expensive to use. A nickel was the price of half a pound of butter in 1892; not many working-class families could afford even one telephone call at that cost, except in extreme emergencies. As well, since the only service available was public telephones, a member of the family had to leave the house to make the call. A mother with young children would have to either leave them alone for a period of time – telephone calls took several minutes to connect, and the caller had to cover a certain distance to get a telephone – or bring them with her, which, in cold weather, entailed dressing them in winter clothes. All of this meant a considerable delay in an emergency situation. Thus, public telephones were underused by these people.

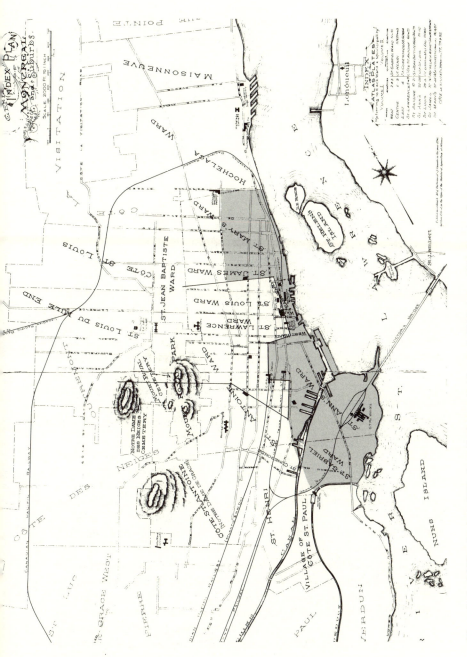

Figure 3: Social Distribution of Telephone Exchanges in Montreal, 1903. (MMA 1881–91. Goad, nos. 1–2, plate 50; BCA, bs 84467, 1561, adaptation of map of "Manual 18,000 telephones 1905")

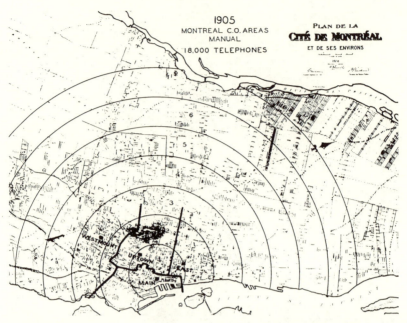

Figure 4: Bell Telephone System in Montreal, 1905. (BCA, sb 84467, 1561, 1905)

Bell Telephone Co. produced its own interpretation of this situation. It stated that the working classes did not understand the utility of the telephone and did not need the service. The proof was that although the company had generously put public telephones at their disposal, working-class families did not use them. The company's explanation can be understood in two ways: it was afraid that supplying telephone service to these groups at low rates would create a precedent that would have negative consequences on the whole system – subscribers asking for lower prices for other types of service – and/or the ruling classes, to which Bell management belonged, thought that it would be "dangerous" to supply these populous and "infamous" areas with a means of communication that could allow them "to gather on the lines" to plot resistance against their exploitation. This might seem far-fetched, but fear of working-class consciousness was common among the dominant classes throughout this period. Attali and Stourdze (1977) report that Louis-Philippe of France had limited access to telegraph lines to very small, approved groups because it was "too political" an instrument to be available to the whole community. As recently as 1918, the United States government took over the telephone system because it considered it too important an instrument of communication to be left in private hands in wartime (BCA, d 18919, 1963, 75).

It is impossible to obtain information on telephone distribution in

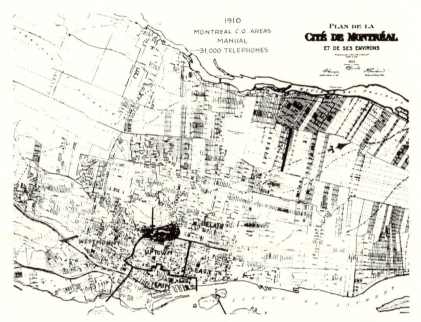

Figure 5: Bell Telephone System in Montreal, 1910. (BCA, sb 84467, 1561, 1910)

Montreal after 1905 from the directories because of the growing number of subscribers. However, some maps do show telephone expansion for the years 1905, 1910, 1915, and 1920 (see figures 4, 5, 6, and 7).

In 1905, there was still only one exchange to serve the most populous area of Montreal (see figure 7),[9] while there were three exchanges in the sparsely populated western sector and one in the centre of the city. In 1909, a new exchange, called St Louis, was opened in central Montreal. It served areas in the northern parts of St Louis and St Lawrence and a suburb north of Mount Royal, called Outremont, mostly occupied by the Francophone middle class and a few middle-class Anglophones. Bell management "had faith" in the expansion of telephone service in this new district because it was home to "30,000 people, made up chiefly of prosperous salaried classes ... The influx of such a desirable population has led to the establishment of one excellent class of stores of all lines, and the district is a rapidly developing commercial centre" (BCA, ncm 1909b). Such social groups, of course, lent "great promise" to the future of the telephone. However, it was not until 1915 that there was noticeable expansion toward the east, with new telephone lines and two addition exchanges. Yet, as a series of articles in *La Presse* revealed, this did not mean that the use of an individual telephone was extended to all classes (BCA, ncm, 1918e, 1918f, 1918g, 1918g, 1918i).

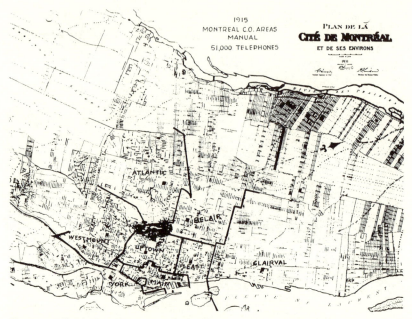

Figure 6: Bell Telephone System in Montreal, 1915. (BCA, sb 84467, 1561, 1915)

The telephone system developed along similar lines in Toronto. In 1894, there were four exchanges: Main, Yorkville, Parkdale, and Junction. Two of them served the business areas, while the other two were in wealthy areas such as Rosedale. By 1915, there were 57,304 stations in Toronto, 47 per cent of which were business connections. Only 6.3 per cent of the total number of residences in Toronto had a telephone. In 1922, the percentage of residential telephones had risen to 10.4 (see figures 8 and 9).

In 1915, there were 51,201 stations in Montreal, half of which were business connections. Thus, only 3.6 per cent of the residences in Montreal had a telephone. By 1922, that percentage had increased to 5.6 (BCA, sb 86245, 1587, 1915, 1922). These numbers are eloquent. They show that even in 1922, the high price of the telephone strictly limited its accessibility to dominant-class and some middle-class households.

The other classes, in both Montreal and Toronto, were said to be served by exchanges with public telephones. In 1883, there were nine public telephones in Montreal (BCA, tdm 1883), eight of them in the business districts. By 1899, there were ninety-seven public stations in Toronto. Seventy of them were in drugstores, and the rest were in various types of stores (BCA, tdm 1899). Drugstores were chosen for public telephones "because the store was opened longer hours than others, as

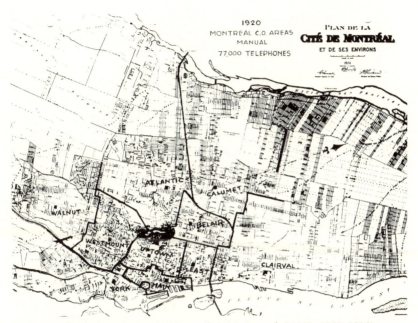

Figure 7: Bell Telephone System in Montreal, 1920. (BCA, sb 84467, 1561, 1920)

well as Sundays and holidays. Furthermore, the druggist was a highly educated man in the community" (BCA, bb 1959). The fact that it took a "highly educated" man to have a public telephone in his store used by working-class families revealed the company's opinion that even this type of telephone was really intended for the educated classes. In reality, book keeping for public telephones was quite simple and did not require a high degree of education. However, working-class people may have expected to find such a highly educated person intimidating. Moreover, public telephones, except for those situated in wealthy environments such as plush hotels, were not enclosed. Since the technology obliged people to shout into the transmitter to be heard at the other end of the line, everyone in the store knew the purpose of the call.

In short, capitalist industry's control over development of the telephone limited its accessibility along class lines. Private interests did not serve the collective interest. In 1922, 90 per cent of the communities "served" by Bell's telephone system were limited to partial access to it – mainly due to its cost, which they could afford only in emergencies – whereas less than 10 per cent of the communities had one or more private lines in their households. Bell refused to provide working-class areas with inexpensive telephone service, despite the fact that they had to run cable through these sectors to reach the ruling-class areas. It seems that political factors were sometimes more important than eco-

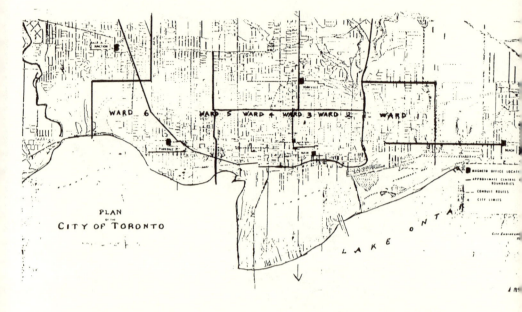

Figure 8: Bell Telephone System in Toronto, circa 1904. (BCA, d 18595, circa 1904)

nomic elements in the development of the telephone. Moreover, as we saw in earlier chapters, users of the telephone system were expected to comply with the dominant moral order, and working-class morality was often questioned by dominant and middle classes.[10] Thus, the telephone system did not expand "naturally." Its geographical and social distribution was influenced by the social conditions under which it developed and, in turn, this distribution affected the types of its use. New social practices emerged.

FROM BUSINESS TO SOCIAL PRACTICES

In 1878, a circular issued by the District Telegraph Co. in Montreal offered free messenger service to any of its subscribers who wanted to reach a non-subscribing doctor, tradesman, plumber, coal dealer, livery-stable keeper, boat builder, florist, or other businessman. Further, "arrangements [could] probably be made for repeating the fire alarm to those who desire[d] it" (BCA, sf.a 1878). The following year, more services were added: access to courts, law offices, banks, druggists, public buildings, hospitals, railways stations, steamboat offices, police stations, and so on. All of these establishments were connected to the central telephone office and, although the subscribers could not

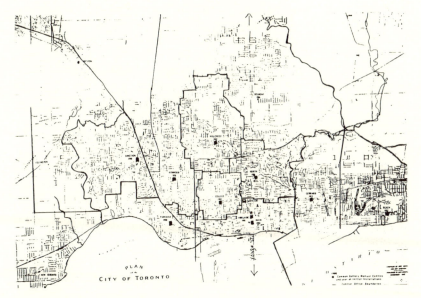

Figure 9: Bell Telephone System in Toronto, 1924. (BCA, d 18597, 1924)

communicate with them directly, the operator could deliver the message. A subscriber could, however, have direct contact, "without leaving the home," with physicians, grocers, meat and fish markets, livery
stables, and so on, which were on the same telephone line. The Dominion Telegraph Co., which was later to become Bell Telephone Co.,
announced that its office was open day and night, and that any of
these places could be reached at any time (BCA, d 12016, 1879c, 1).
That same year, the company advertised in newspapers that subscribers
"can be placed in direct and 'instant communication' [with] prominent
members of the legal, medical and other professions" (BCA, d 29920,
1879).

However, doctors and lawyers were not immediately drawn to the
telephone. Lawyers feared a lack of privacy. In Montreal, "the Honor-
able Mr Abbott, who afterward was Prime Minister, wouldn't have a
telephone in his office because it was beneath the dignity of a lawyer
to communicate with his clients over such a contraption" (BCA, sf.ind
1931). Another lawyer was said to have had a hole drilled in his vault
so that he could use his telephone in a place where he thought he could
have complete privacy for his business transactions (BCA, qa 1979). The
question of telephone "trust" was acute at this time. The early technology did not offer the degree of "secrecy" to which late-Victorian
bourgeois and petty-bourgeois classes were used.

Doctors did not trust the telephone any more than did lawyers, but for a different reason: "They were afraid they would never be able to collect bills due for advice given away from the presence of the patient" (BCA, sf.ind 1931). While the lawyers overcame their fear of indiscretion with private lines, the physicians' concern was not addressed, since their lack of trust was not in the technology itself, but in its use by their patients. Bell considered access to doctors through the telephone to be essential to enhancing its attractiveness to consumers. To get doctors to subscribe, the company offered them special rates.[11] In January of 1880, a proposal to connect the doctors of Montreal to Bell's exchange was studied by the Medico-Surgical Society in a general meeting, and "the resolution accepting it was unanimously carried" (BCA, d 12016, 1880c, 18). Soon after, the telephone came to be considered "a necessary part of the doctor's equipment." It was "compared with a long stethoscope with which the physician keeps in touch with the needs of his patients." Nevertheless, its invasive character remained a serious drawback for the doctor whom it brought "into communication with the world when he would not, as well as when he would have it so" ("The Telephone and the Doctor," 1912, 639).

Monopolizing medical services on its telephone lines gave Bell Telephone Co. a great advantage over other telephone companies in Montreal. A Montreal newspaper accurately stressed that "it secure[d] them the patronage of all parties desiring telephonic communication with their family physicians, and [was] an advantage to the doctors in placing them within a moment's call of each other in case of consultation" (BCA, ncm 1880a).

This "big stroke," as the newspaper called it, enabled Bell to defeat its main competitors. In Montreal in the late 1870s, there were two companies – the District Telegraph Co., which used Edison's telephone, and the Dominion Telegraph Co., which used Bell's – operating completely independently. This meant that subscribers to each company could not communicate with one another unless they subscribed to both companies. Since the medical community had been secured by Bell users of the other company could not reach any of these doctors by means of their telephone. They had to either use the telegraph, which took much longer to transmit a message, or change telephone companies. Several of them chose the latter option. Bell's coup was so significant that, just a few weeks later, the District Telegraph Co. decided to sell out to Bell.

No documents remain to show that a similar agreement was reached with doctors in Toronto. However, in general, the same services were offered in Toronto as in Montreal. For instance, telephone subscribers in Toronto also had a messenger service, with "neatly uniformed messengers," which supplemented telephone service by delivering letters and parcels (BCA, d 3653, 1882).

Another important development in terms of "social services" was when luxury hotels started to put a telephone in each bedroom. In Toronto, the very first telephone was placed in the Queen's Hotel (BCA, bp 1980). In 1902, the Windsor Hotel in Montreal ordered four hundred telephones for installation in each guest room and each department. It was the first Canadian hotel to provide such extensive telephone service to its guests "free of cost" (BCA, ncm 1902b). Other hotels had to follow suit to remain competitive. Telephones in hotel bedrooms were used for outside as well as inside connections. Now, room service and information were instantaneously available simply by using the telephone. This may have changed the leisurely pace of life for guests who were travelling for pleasure (as some claimed), but it accelerated delivery for the impatient businessmen at whom the service was aimed. On the other hand, the use of telephones in hotels – and in other businesses – contributed to a reduction in the number of messengers, and threatened the existence of the occupation. The use of telephones in restaurants came later. In 1915, "first class restaurants and tea rooms," particularly in American cities, tried to replace waitresses with telephones which "permit[ted] the guests to communicate their orders direct to the kitchen without suffering delay" (BCA, ncm 1915a). Diners could also communicate by telephone between tables, a practice that seems to have been very popular with wealthy women (BCA, ty 10 (3) 1905, 211). However, this use of the telephone was considered frivolous, and the service was discontinued, although, in some ways, it was a labour-saving device.

Railway stations were slow to adopt the telephone. In the early 1880s, telephones were used at the ticket counter and in other departments to give out information on train arrivals, departures, and rates, but there was no system connecting railway stations. Railways had traditionally used the telegraph. Station managers argued that the telephone was unreliable and that they could not afford mistakes. In 1906, however, the Grand Trunk decided "to establish a long-distance telephone circuit over its whole system ... as the instrument and medium of control ... of the terminus and movement of cars from station to station," since, "fast as the telegraph seems to be, it is not fast enough to meet the case." As it was, trains were kept "idle" because of the slowness of the telegraph, and merchandise waited in railcars. The telephone was felt to be more expeditious and safer (BCA, ncm 1906b). According to *Scientific American*, the advantage of the telephone over the telegraph were that personal orders via the latter medium ensured better comprehension; the telephone was more rapid, thus making the railway system more productive; portable telephone sets allowed conductors to be in immediate contact with dispatchers, which was safer; and the telephone permitted a greater accuracy in transmitting orders, which was more reliable ("The Telephone and Railways," 1910, 388). All of these elements converged toward a single goal: faster communication for faster circulation and, thus, faster accumulation of capital. The adoption of the telephone by the railway system to accelerate the transportation of commodities confirms that the development of faster means of communication in industrial society is supported by capital inasmuch as it reduces the time of circulation and increases accumulation.

In fact, almost all of the services available via the telephone were more or less related to ruling-class activities. Of course, the telephone provided access to such public services as the fire department, the police, and doctors, but all of these could already be reached by strategically located telegraph boxes, which for working-class families was as least as fast as the telephone. All of the "social services" provided by the telephone system were more rapid and efficient only for those who had a station in their homes; even as late as 1922, only about 6 per cent of the households in Toronto and Montreal contained a telephone. These telephone services did not improve the life of the majority. With the exception of a few charity cases subsidized by Bell,[12] the domestic telephone remained far too expensive for the working-class household.

The degree of access to the telephone differed according to social class: there were individual telephones in wealthy households, and public telephones for working-class families. This access, however, did not give users entire control over the mode of consumption of the service. Although subscribers had some influence in determination of telephone uses, there was often a discrepancy between actual and desired uses. Among the factors responsible for this discrepancy was the existence of gender-oriented cultural practices specific to late-Victorian society.

TRANSITION: GETTING FAMILIAR WITH THE TELEPHONE

In Canada, the telephone system controlled by Bell Telephone Co. developed mainly among the ruling classes in cities and towns. The members of Bell's board of directors were drawn from the high bourgeoisie ("haute bourgeoisie"), a small group of English and Scottish businessmen and a few French-Canadians, who owned masses of capital, controlled major businesses, and shaped political institutions. The Canadian high bourgeoisie of the late nineteenth century lived almost exclusively in cities, and had six major sources of capital accumulation: import-export trade, commercial credit, real-estate speculation, water and railway transportation, large public works, and industrial production (Linteau, Durocher and Robert 1983, 142). Members of this group had started out as agents or partners of English and Scottish merchants, although as late as 1885 a good percentage of the Canadian industrial elite consisted of the sons of farmers or craftsmen. Representation by the latter declined rapidly over the years, however, and by the early twentieth century the ruling classes were largely endogamous. A less influential but larger group, the bourgeoisie, whose members were English and Scottish with a significant number of Irish and French-Canadians, possessed smaller amounts of capital. Their political and economic power was concentrated in medium-sized regional institutions. Although some of them lived in large cities, a large group comprised the ruling class in middle-sized towns (Linteau, Durocher, & Robert 1983).

The petty bourgeoisie consisted partly of small entrepreneurs and retail merchants with a small amount of capital who employed only a few workers, and partly of professionals. This class had limited economic influence, and its political power was exercised mostly in Parliament, city and town councils, and school boards, where it usually followed the lead of the bourgeoisie. Petty-bourgeois influence was more noticeable at the parish and neighbourhood levels. However, the most characteristic manifestation of petty-bourgeoisie power occurred in small towns and rural areas, where its members occupied administrative positions in various institutions.

Economically and politically, these classes can be situated clearly; the culture of class relations is more problematic. A strict etiquette rules social contacts between classes. A petty bourgeois, for instance, needed a special invitation to enter the house of a high bourgeois, whereas the latter could casually drop into the former's home.

Women typically had the social status of their husbands, although they did not enjoy their power. Nonetheless, in the household, they had control over servants and children. Most bourgeois women, for example, employed several maids. In Montreal, 66 per cent of the total number of domestic servants worked for bourgeois English families living in St Antoine. Some petty-bourgeois households also had maids (Lavigne and Pinard 1977, 37). Since they could afford domestic help, these women had daily periods of leisure time. Their tasks, mainly limited to managing domestic chores, were usually completed in the morning. Petty-bourgeois women generally went to the market to shop for food, with the auxiliary motive of meeting friends (Katz 1975, 4).

Their afternoons were devoted to volunteer work in various associations, mostly religious, or to visiting friends or relatives (Lavigne and Pinard 1977, 41).

Many women participated in volunteer associations which "performed a multitude of community functions: care of the poor, instruction of children, self-improvement of young men and simple conviviality for their members" (Ryan 1983, 169). Ryan asserts that women belonging to these associations were linked through large networks which provided them with support from other women that reached beyond household affairs, and "extended to a rich social network based in the long-term, habitual cooperation of women in church and reform activities" (Ryan 1983, 181). As such, their influence reached various social spheres, often beyond the immediate community to the regional and even the national levels. As well, a small group of economically independent middle-class women were busy owning and managing boarding houses, grocery stores, and other small businesses (Lavigne and Pinard 1977, 51).

In the late nineteenth and early twentieth centuries, it was not uncommon to find petty-bourgeois women in rural areas who had very limited financial means. Their situation obliged these women to do domestic chores such as sewing, gardening, and so on, while they hired a maid to maintain the house (Fahmy-Eid and Dumont 1983, 251). Farmers' wives, on the other hand, had long work days during which they had to do everything in and around the house, including milking the cows and feeding the poultry, pigs, and other livestock. This had an important impact on their social life: they did not have any spare time for socializing during the week. Family visits to relatives and friends were strictly limited to Sunday afternoons (Miner 1939, 144), while close male friends or relatives often dropped in during the evening for a smoke (Miner 1939, 143). In contrast, the patterns of sociability of the bourgeois and petty-bourgeois classes were more formally defined, and very much based on a particular concept of privacy.

In Montreal in the late 1870s and early 1880s, a city of wooden sidewalks,[13] horse-drawn tramways, and gas-lighted street lamps (MMA 1947), the entertainments of the dominant and upper-middle classes were in the form of plays, concerts, balls, and so on (all of which were prohibited by the church).[14] Festivals were another occasion for socializing (Katz 1975, 3). However, visiting was the most important form of recreation, and it was strictly circumscribed by rules. Mornings were inappropriate for visits, except for close friends or relatives, or on very special occasions; this was the time to make invitations for tea, dinner, or evening entertainment. Before the advent of the telephone, invitations were made in various ways. The most casual ones were made in per-

son.[15] Someone would drop by in the morning and invite the family –
or the parents only – to a reception given by another family, usually
from the same class. Invitations were also made when people met at the
market. Generally, though, they were delivered through notes brought
by a servant, or by telegrams. For both, a formal response was expected.

In the 1870s in Montreal, and in all large Canadian cities, very complete telegraph networks existed. Telegraph boxes were located in various spots around the city. To use them, people "repeat[ed] [the] signals
in the manner indicated in the printed instructions" (BCA, d 6454, nd).
Some telegraph boxes were privately owned, as telephones were later.
Several wealthy households possessed such a box, and could thereby
communicate with telegraph offices, police stations, fire departments,
grocers, drugstores, coal merchants, and so on. Storekeepers paid a fee
to have their names printed on the boxes, and to have a private box
directly connected to the network. The procedure was as follows: when
a housewife wanted to order some meat from a butcher, for example,
she pushed a button marked "butcher." The call was registered by the
telegraph central office, with which all boxes were connected. The office
sent a message to the butcher shop, and the butcher sent a messenger
to the house to take the order. In households that did not have a private
telegraph box, the housekeeper could send a maid to the corner street
box to make the call. These practices were very common in cities before
telephone service started to expand.[16] In fact, many of the first telephone
subscribers already had a telegraph box.

Thus, when the telephone appeared, some members of the ruling
classes were already acquainted with long-distance means of communication. Nonetheless, telegraph practices and telephone practices were
quite different. In one magazine, a writer compared the telegraph to a
page of print and the telephone to an oil painting:

A telephonic message differs as widely from a telegraphic message as a highly
finished oil-painting differs from a page of print. In the one you have only black
and white, black symbols on a white ground, the symbols being limited in
number, and recurring again and again with mere differences of order. The
painting, on the other hand, discloses every variety of colour and arrangement
... the tints shade off gradually and softly into each other, presenting tone and
depth in endless variety. The page of print is unintelligeable [sic] without the
aid of the key; the painting tells its story plainly enough to anyone who had
eyes to see. ("Introduction of the Telephone" 1878, 208)

Learning to use the telephone was not easy for the first generation of
subscribers. According to the telephone companies, subscribers had to
be "educated."

EDUCATING TELEPHONE USERS

The telephone system started to expand during a period of remarkable technological development. Whereas the telegraph had been in use for more than twenty years without any serious competition from other technological inventions, the telephone came out among a group of other technologies which sometimes were detrimental to its operation. In Montreal, for instance, after the telephone became commercially available in 1878, electric street lights were installed in 1884, the electric tramway appeared in 1892, and the wireless telegraph came in 1901 (MMA 1942b). A writer in *Bell Telephone Quarterly* explained that only the "frantic" way of life of that time could have led to the development of the telephone. "With the faster tempo of industry and commerce, the time was soon to come when a nation geared to high speed would no longer be satisfied to make its business and social contacts by means of one-way communication, no matter how rapid it was. The needs of a nation demand some form of communication that could overcome distance and time, but that would be, nevertheless, as intimate and personal, as direct and reciprocal, as a face-to-face conversation. That form of communication, only the telephone, by transmitting the voice itself, could provide" (Barrett 1940, 129–38). In spite of its apparent simplicity, however, the use of this "savior of the nation" required the learning of some skills.

Newspaper reports and advertisements, along with articles in magazines and journals, often stressed that use of the telephone required no skills or training. Yet the same media repeatedly emphasized that people needed to be "educated" in the "art of telephoning." This education was conferred in the form of instructions and rules published in telephone directories. Starting in the early 1880s, and for about ten years thereafter, instructions on how to use the telephone were printed at the bottom of almost every page, and were repeated several times elsewhere in telephone books. In the 1883 Montreal directory, for instance, fourteen instructions were repeated throughout the fifty-one pages of the book. The most important were reproduced more often – for example "Answer calls promptly" was printed five times, whereas "Write names of new subscribers in this book as received" recurred only three times. The instructions covered several other aspects of telephone use. For example, when requesting a connection, users were asked to pronounce the name[17] of the party wanted with "especial distinction" (MMA 1883, 13). When the connection was made, the parties were instructed to "speak into the transmitter, with the mouth six or eight inches from it, speak distinctly and somewhat slowly" (MMA 1883, 11). People were also requested to hang up their telephones properly

(MMA 1883, 4), and not to take them off the hook "unless for purpose of conversation" (MMA 1883, 10). According to Bell Telephone Co., leaving the receiver off the hook "destroy[ed] the feeling of confidence in privacy of the service ... and result[ed] in unnecessary expense to the company, as the line [was] reported in trouble without cause" (BCA, qa, ndb, 118). When it was necessary to interrupt a conversation temporarily, people were told to "keep the telephone at ear" (MMA 1883, 5). They were also instructed to handle the phone carefully (MMA 1883, 15), and to speak into the receiver should the transmitter be broken (MMA 1883, 9). Subscribers were warned that they should "not attempt to use the telephone on the approach of, or during, a thunder storm," because of the risk of electrical shock (MMA 1883, 7). Finally, they were warned that the instrument was leased *for the use of the subscribers only* (MMA 1883, 8, my emphasis). This first set of instructions reveals that some skills were required to use the telephone, and to understand how it worked. Indeed, these elementary procedures could easily be mastered by anyone *who could read the instructions*, as Bell Telephone Co. apparently expected that its subscribers could and would do. Because the telephone was aimed at literate people, those who could not read could not acquire telephone skills. The apparently universal access to the telephone was thereby limited by the companies in other ways.

A second set of instructions was published in the 1884 Montreal directory. This new edition contained, in addition to the first set of rules, instructions on the operation of the system. Users were notified to "call central office with one ring only" (MMA 1884, 32), and warned that "the operator could not hold conversation with them" (MMA 1884, 15). They were also informed that "complaints to operators avail nothing" (MMA 1884, 15), because their "duty [is] to answer calls only" (MMA 1884, 21). Furthermore, subscribers were to be rung *only twice*. This attempt at disciplining users was intended to accelerate the process of connecting parties, and therefore to decrease unprofitable use of the lines. If there was no answer after two rings, the call was disconnected. (Subscribers who could not read this particular rule may have found it frustrating to use the telephone until they understood the process.) There were also instructions regarding the etiquette of telephone answering, which recommended that answerers give their name and refuse to speak to people who declined to identify themselves.[18] Finally, instructions were also given for calls to fire departments, train stations, police stations, and so on.

In 1890, all subscribers were warned that they were entirely "responsible for all connections made on their telephone over trunk lines" (MMA 1890, 2). A corollary to this new rule was that no message could be sent "collect" (MMA 1890, 3). Bell had chosen this means to discipline subscribers into a behaviour that they were not very inclined to adopt: refusing the use of their telephones to non-subscribers. There were several complaints by Bell managers regarding subscribers' unwillingness to limit the use of their telephone to themselves, and regarding their tendency to "abuse" the company's system (BCA, d 1239, nd). For the company, the offence became an outrage when the telephone had been supplied free of change to the subscriber. This was the case with a hospital in London, Ontario. The institution had been given a free telephone by Bell, which had a phone booth in the vicinity. However, the hospital allowed people to use its apparatus, thus depriving Bell of an income from the public telephone. General manager Sise was furious, and wrote the hospital administrator that he would have the telephone disconnected if the institution did not stop this "abusive" practice (BCA, d 12267, 1884).

In 1899, instructions addressed the voice and elocution of users, telling how to pronounce a telephone number properly (BCA, tdm 1899, 3). This instruction, which was required of the operators, was presented as an element of telephone etiquette to the subscribers. The directions indicated very precisely what the subscribers "should do" and "should not do." This aspect of telephone use was considered important because, as an operator stated in a letter to the *Star*, "If half the subscribers would come on the line and repeat their numbers slowly and distinctly, you would get no wrong number" (BCA, ncm 1919e). In this edition of the telephone directory, "instructions" were clearly separated from the "rules and regulations"; heretofore, both sets of notices had been published under one title, "Notices to Subscribers." Now, however, all notices were grouped under "Rules and Regulations" except instructions for technical use of the telephone, which were titled "General Instructions." Below this title was printed, in capital letters, with a finger pointing toward the message, "Destroy previous list."

Naturally, the instructions for use of the telephone changed over the years, as the apparatus improved. For instance, with the advent of common-battery switchboard, users did not have to ring the operator anymore, since a light automatically went on when the subscribers took their receivers off the hook. Other modifications occurred when users had to learn how to dial the automatic telephone. Automation was seen as the ultimate development in telephony, and much was at stake. For this reason, Bell Telephone Co. sent six men to New York to learn how to handle automatic telephones and, "on their return to Toronto, an educational campaign was started." This campaign was directed not only at subscribers, but also at the "general public." It consisted of Bell employees training certain people, through personal contact, using "personal letters and interviews, demonstrations, moving pictures, informative booklets and folders ... Practically every public body, and particularly the branches of the public service, the police, firemen, etc., have been given special instructions." The campaign was institution- alized by its insertion into the school curriculum, with "talks [given] before school children," who were told to teach their parents (BCA, bb 1924, 15). For this important step, the company left nothing to chance.

In addition to these explicit instructions, *implicit* rules were often suggested either in newspaper reports or in journal articles, including how the subscriber should proceed if he was moving (BCA, ncm 1905c), and how to have one's new telephone number printed in the upcoming directory (BCA, ncm 1908d). Newspaper reports and magazine articles were also an essential means of transmittal of Bell's messages to sub- scribers, for instance, informing them of their responsibilities in making successful telephone calls, and asking for their co-operation (BCA, ncm 1909a). In *Bell Telephone Quarterly*, K.W. Waterson summarized the subscribers' elements of co-operation in six points: "answer a call promptly and courteously"; "have sufficient telephone facilities so that one's lines [are] not ... busy an undue portion of time"; "make provision for someone else to answer it [the telephone] properly" if the subscriber did not want to do so himself; "know how to make ... calls ... give the call accurately and clearly to the operator ... pay attention to her rep- etition"; "know how to use the instrument ... the significance of signals ... and how to signal the operator"; and on party lines, refrain from interference with other persons on the lines (Waterson 1922, 26). These points implied that subscribers knew how many telephones to buy, the right way to ask for and answer a connection, and of good manners approved by the users from the ruling classes. To ensure subscribers' co-operation and to make them understand their "responsibility," news- papers published a series of "announcements" produced by Bell that explained the functioning of the telephone system, thereby educating the users about what they ought to know (that is, what Bell wanted them to know). The official reason for such publications was to give users "a better idea of the human, personal side of the telephone system" (BCA, d 1978, 1920). It was in fact a way of using subscribers' sense of decency to reduce unprofitable consumption and increase the rate of production of telephone calls.

Party lines, which were the majority of lines at the time, particularly required the co-operation of subscribers for their successful operation. Telephone companies published lists of rules for party-line users, stress- ing the most frequent "abuses and annoyances" and "their remedies." One of the most persistent "abuses" was "lengthy 'visiting' conversa- tions, lasting for half-an-hour or more" (BCA, d 29516, nd; sf.ind 1904). These conversations, usually involving women, were considered "frivolous," and the "real" motive for the call was thought to necessitate no more than two minutes on the line: "It is, of course, well understood that business conversations cannot be limited as to time, but 'visiting' can beneficially be confined to reasonable short duration of time" (BCA, d 29516., nd). Another "unneighbourly abuse" was "the habit of removing the receiver on all signals, and listening in on other people's conversation." This practice reduced transmission strength and made hearing more difficult for the two parties in connection. The "remedy," of course, was "*to answer your own signal only*" (original emphasis). A third source of annoyance was children playing with the telephone, "taking down the receiver and shouting into the transmitter." The solution to this problem was for families with "good" parents to "instruct the children[19] on what to do, and what not do with the telephone" (BCA, d 29516, nd). In other words, the parents ought to pass onto their children what they themselves had been instructed by the telephone company.

This co-operation came to be officially associated with a new concept in social practices: "telephone etiquette." "Courtesy, consideration, and common sense [were] the prime requisites of telephone etiquette" (BCA, d 29547, nd). Bell Telephone Co. "insist[ed] upon a courteous, considerate and obliging attitude" from both subscribers and workers. "The use of profane or improper language over the lines [was] strictly forbidden." According to C.F. Sise, "necessity" had forced the company to lay down "rigid rules," and these ought to be adhered to (BCA, d 1857, 1916). Courtesy on the telephone implied several things, but at the top of the list was considerate language. Telephone etiquette was to be applied during the entire process of telephone calling, from answering to ending a connection. It was related not only to the ruling classes' moral values, but to popular culture as well, and thus changed over the years. For instance, the taken-for-granted "Hello" of nowadays was considered impolite in early days. "Well?" or "Are you there?" were more appropriate in late-Victorian society (BCA, me, 1880). Moreover, when users were connected to a wrong number, they were expected to use the very formal "I beg your pardon." It was also "good manners" to offer to pay for the call (even a local call) when telephoning from a friend's or an acquaintance's house; under these circumstances one should limit oneself to what had to be said, and not tie up the line with "aimless chatter" (BCA, ncm 1915a). These were dominant- and middle-class manners. Of course, not all ruling-class users were courteous all the time. The technology was not always reliable, and users had to be patient. Those who became impatient and did not follow the rules were disciplined.

Discipline was imposed in several ways. Discourteous users were "cut off," and very serious offenders sometimes had their telephones removed. This, however, was the last resort, because it was "not good policy as one man [might] sometimes do [the telephone company's] business great injury when opposed to it" (BCA, qa 1881b, 21). Notwithstanding the companies' reluctance to remove a station, this dramatic step was taken from time to time (BCA, ls 1910–13, 5). In fact, disciplining was so efficient that, in 1935, a manager could assert that telephone companies had finally succeeded in "forc[ing] users *against their will* to bend their normal actions to fit their communication tools, rather than the reverse" (Jewett 1935, 184, my emphasis).

In spite of rules and discipline, some individuals continued to use the telephone incorrectly. Some "drop[ped] the receiver and describe[d] with gestures the dimensions or shape of some articles [they were] discussing" (BCA, ls 1889). Others thought that the telephone was a living entity and talked *to* the instrument rather than *into* it. The companies claimed that women were particularly inclined to do this, for instance, asking the telephone to wait when it rang because they were busy with something else (BCA, sf.ind 1900, 22). There were also those who did not bother to inquire who had answered and immediately started shouting idiosyncratic language into the transmitter, to the shock of the stranger of the other end of the line (BCA, qa 1884b, 69). Finally, some used their telephones as convenient shelves rather than as a means of communication, placing objects on the instrument which sometimes short-circuited the entire line (BCA, qa 1880b, 83). Claims about misuse of the telephone were also made regarding entire groups. For instance, French-Canadians were said to stubbornly deny the utility of the telephone altogether (Fetherstonhaugh 1944, 118). Women were considered by the company to be incorrigible "delinquents" – talking for hours and monopolizing telephone lines which could be better used for business. A reporter for the *Globe and Mail* went so far as to ask if women should "be permitted to use the telephone without obtaining a permit certifying to their competence to handle the instrument" (BCA, sf.wo 1951). As I shall show, women were the victim of sexist and sarcastic jokes concerning the telephone. Finally, some citizens thought that telephone wires were hollow and carried bacteria wich were responsible for the 1885 smallpox epidemic. Some people were so angry about the telephone as an agent of disease that they marched to the telephone exchange "with torches, pick-axes and clubs when, *luckily*, a group of militia on its weekly parade came up the street just in time to avert what could have been a disaster" (BCA, me 7b 1885, my emphasis).

In short, the use of this new technology, which apparently required no skills, actually involved the learning of a set of instructions and rules. These tended to limit its accessibility to a specific group of people. It was expected that users would co-operate to ensure the success of the system, and those who did not comply were disciplined. In spite of this, some groups and individuals continued to misuse the telephone. The phone was regularly used by subscribers in ways which seem unusual to the contemporary reader.

EARLY USES OF THE TELEPHONE

Early uses of the telephone differed from contemporary uses, partly due to the technology itself, and partly due to other factors. For instance, the fact that transportation was often difficult, especially in winter and spring, accounted for certain specific uses of the telephone. Another important element was that, before 1920, commercial radio did not exist in Canada. Although wireless telegraph transmission was available, it was not used for mass communications. The telephone was the technology used for this purpose. Since telephone networks were organized in "domestic circuits" (BCA, sf.id 1877) – several households and businesses connected to a collective line – the telephone constituted a form of group communication which led to the creation of activities that sometimes resembled radiophony. In 1880, with the opening of the first exchange, the formal structure of the telephone system changed, and the "domestic circuits" were connected to a central office. This did not modify the collective characteristic of the network; rather, it contributed to its extension, so that when a communication was made through the exchange all subscribers could hear it.

Business telephones were dominated by a rational use: capital accumulation. However, the "rational" use of a residential telephone, beyond linking the businessman's office to his home and the housewife to her suppliers, was a matter of debate. One of the first long-distance uses of residential telephony was to "bring God to the home of the sick and invalid." In 1879, the Dominion Telegraph Co. of Montreal allowed an invalid to hear the mass and the priest's sermon from a church that was "fully a mile and a half" from his residence. The *Herald* reported that the transmission was "characterised with remarkable distinctness" (BCA, qa 1879). An experiment by the rival company, Bell Telephone, took the form of a church service transmitted to "hotels, public buildings, and business offices, and private residences of our leading citizens" (BCA, qa 1880a). Several other church activities were transmitted by telephone in Toronto, Hamilton, Ottawa, Quebec City, and other cities.[20] As transmission distance increased, telephone companies started to provide church services to summer houses equipped with a telephone. For instance, "Montrealers at Highgate Spring [took] part in the service in the St James Street Methodist Church." This was a distance of ninety-two miles, and the experiment was said to be very successful, as the parishoners of that church "residing at the charming summer resort were enabled to take part in the devotions as perfectly as if they were in the church itself" (BCA, d 12015, 1881). A moral issue was thereby raised: was a religious service over the telephone as sacred as one at church? While the answer to this question remained problematic, there was no doubt that political discourses were considered to be as efficient when transmitted by telephone as when delivered before a crowd. Besided, it was a way to reach a larger number of people.[21]

The phone was also used for entertainment. Concerts, recitations, poetry readings, and so on, started to be transmitted by telephone very early in its development. In fact, it was A.G. Bell who, to demonstrate the efficiency of this invention, asked some singers, actors, and so on, to perform over the apparatus (BCA, d 12015, 1877a, 1). Hence, when the domestic circuits were constructed, various forms of entertainment were organized within groups of subscribers. Later, these concerts were extended to all users in an exchange (BCA, ty 8 (3) 1904, 232).

The first cultural activities on the telephone occurred in 1879 on the exchanges joining Brantford, Hamilton, and Toronto. They included recitations and readings of literary works "over a line 100 miles long." Although people on the line were not acquainted and were separated by a large distance, they reported that they had experienced a "feeling of closeness" among them (BCA, qa 1877). In 1880, concerts given in Windsor, Ontario, were transmitted to Detroit and Sandwich. The reception was said to be "excellent" (BCA, d 12016, 1880f). In Ottawa, in 1881, a concert by the Governor-General's Foot Guards was "broad-casted" by Bell Telephone to seventy telephone subscribers "at one time." "The performance varied from half a mile to thirteen miles, and in every case the effect was excellent" (BCA, d 12016, 1881). One of the most successful telephonic entertainment originated in Kingston and Belleville, and was called "Sunday Evening Musicales." Each Sunday evening, subscribers could listen to a concert over their telephones. When the organizers were short of performers, employees were asked to perform (BCA, d 12015, 1877a). This medium of entertainment were successful, although some listeners remarked that the effect was very different from that obtained when sitting in a concert hall or theatre: "As yet each individual requires an instrument placed close to the ear in order to hear the distant sound, as children put shells to their ears to hear the sound of the sea" (BCA, d 12015, 1877b). But transmission of artistic performances was not encouraged by Bell Telephone Co.; it was considered an abuse of the lines, since it monopolized them for long periods. In a letter to W.J. Gilmour, a manager in Ontario, L.B. McFarlane warned him to force subscribers to stop "amus[ing] themselves in various ways, [and] listening to music over lines, etc.," or the company would refuse to repair the lines, and it would become "impossible to keep their service in order for commercial uses" (BCA, sf.ind 1904). McFarlane had pushed the right button: making profit through business transactions over the phone. In any case, the idea of listening to a musical performance from a distance was a bewildering and dazzling precedent most for people. As a writer put it,

> You can list to a concert and never go out,
> But can hear every song that is sung;
> You can easily know what a play is about
> From the line when the curtain's uprung.
> You can hear the debates in the House if you like,
> But that twaddle might make many telephones strike. (BCA., qa 1880c)

Indeed, "mass transmission" over the telephone was also used for other purposes, including as a "speaking newspaper." A long-lasting instance of newspaper transmission by telephone was the "Telefon-Hirmondo Telephone Newspaper," in Budapest, which began in 1893 and lasted for over eighteen years. It concentrated on news, and the handling of the information was said to be quite sophisticated and rapid. For instance, "special flashes" were transmitted to the subscribers' lines before they could read them in the newspapers (Briggs 1977, 52–3). In Newmark, New Jersey, subscribers connected to a special circuit said that they received the news of the day more promptly than with other means of communication. This telephonic newspaper was called "The Telephone Herald." It existed in the form of one continuous edition from 8:00 a.m. to 10:30 p.m., and cost five cents per day. This elaborate information program lasted about a year.[22]

However, other types of information systems, developed on a much smaller scale and limited to local distribution over the telephone, were maintained by the operators. Actually, the central exchange was considered "a sort of local newspaper office, receiving and dispensing news and information" (BCA, d 29909, 1967, 41). In a small town without a daily newspaper, for instance, each morning an operator "advertised by ringing eight longs and then making general announcements – special on groceries, current ... shows, and the Missionary Society's ice-cream social. She gave out bulletins on the sick. She was a clearing-house for details of funeral arrangements" (BCA, ls 1880–91, 1). "Like a newspaper office, the telephone exchange ha[d] come to be considered by the great world as the compendium of every item of useful information in the world" (BCA, ty 9 (5) 1905, 389). Operators' testimonies reveal that they gave out all kinds of information, from the name of a butcher to football scores. In most exchanges, there were more or less formally organized systems through which telephone operators gave updates of returns on election days. Telephonic transmission of information, however, did not encourage mass communication, but merely bound together specific social groups.

Other types of service were also offered by various groups. In some communities, butchers, grocers, and other storekeepers called their customers in the morning to take their orders (BCA, d 12444, 1952). Storekeepers who did not have a phone often had an arrangement with the operators, who acted as intermediaries between them and their customers (BCA, d 3379, 1926, 1). Conveniences such as cabs, messengers, delivery of letters, and so on were supplied by telephone companies for a small fee (BCA, bg 1880). Other frequently provided services were wake-up calls so that people could catch early trains (BCA, tdm 1883); weather-forecast transmission, particularly for rural subscribers (BCA, ty 8 (1) 1904, 12); and even mail delivery – the postmaster could read "special letters" on the telephone to people who could not promptly go to the post office (BCA, ty 7 (1) 1904, 71).

The diversity in early uses of the residential telephone was astonishing. Early telephony engendered imaginative uses which have since disappeared. The examination of early telephone practices in relation to political-economic development shows that most of these original and spontaneous uses were later eliminated. Those that remained cannot be attributed solely to the technical attributes of the telephone. The telephone company's desire for increased profit, as well as certain political motives, limited accessibility to the phone to certain classes and, within those classes, to certain uses. As such, the telephone was not only class-oriented, but also gender-oriented, as the recommended uses were mostly adapted to men. Nonetheless, although women were very much neglected in telephone development, they found ways to exploit the little room for action that they were given, and succeeded in significantly influencing the "culture of the telephone." This will be examined in the next chapter.

CHAPTER SIX

The Culture of the Telephone

In the 1890s, women in the wealthy classes were using a telephone system built and shaped by the economic incentives of the telephone industry and by political and ideological forces. Women's activities were not seen as being of prime importance in the business world of the telephone entrepreneurs. Nor did these entrepreneurs see the utility of this new technology for working-class housewives, or for rural populations. As Marvin pointed out, telephone-company managers thought that "women's use of men's technology would come to no good end" (1988, 23). Capitalists considered the telephone only to be a means of facilitating business activity or to link the businessman to his office when he chose to stay at home. Its initial impact on other social groups was slight.

This began to change when the telephonic network expanded to residential areas. This expansion resulted from a developmental dynamic involving several components. No general consensus among women forced telephone companies to extend the telephone system for their use. At first, women's access to the telephone on a personal basis was very much subject to their husbands' trust in the technology. Businessmen who used the telephone system generally had their offices connected to their homes. Consequently, women gained access to the telephone, at first for practices recommended by the companies, but soon for activities of their own. The early structure of telephonic networks shows that they were used primarily within friendship circles, which later expanded as new exchanges were opened. Thus, for some subscribers, the opening of an exchange was the equivalent of the "pedestrianization" of their elite telephone network. It was from within closed and approved circles that the domestic use of the telephone grew, accompanied, much later, by telephone companies' advertising of the utility of the telephone for women's practices. Only when women started to use the telephone extensively for their own activities could a (female) telephone culture emerge. Women's contribution to changing the social practices of telephone use was important, although we must be critical of contemporary male accounts of it.

OBTAINING THE USE OF A TELEPHONE

Means of communication are not determined simply by technology. Siegelaub says that a form of communication represents "a bond between real people taking place in real time and real space" (1979, 12). As a link between people, it is bound to take different forms when applied to different classes, cultures, and epochs. In effect, to situate the practices related to a new form of communication is to associate the terms "communication" and "culture." This "world of reference" varies according to the structure of social classes, and according to the nature of social organization and the degree of people's solidarity in daily life. This results in the development of divergent networks which make possible other uses of the means of communication (Mattelart 1983). This differentiation in the world of reference is also gender-related. M. Mattelart (1985b) asserts that women have different communicational activities from men, and that a means of communication adapted to male practices will not necessarily be suitable to female activities. Hence, women may use a means of communication in ways unexpected by men.

The telephone developed in different ways according to the period and the areas in which the systems expanded. Indeed, technological development of telephone systems changed dramatically over the period studied, and these modifications influenced the production of telephone calls as well as the use of the telephone. It is to be expected, then, that these technological developments had also some impact on the reproduction of social activities. Nevertheless, the period during which expansion took place was not the only factor affecting telephone systems; the areas in which expansion took place were also influential. For instance, an examination of telephone systems reveals that more sophisticated technology was used on more important routes. Use of the system was influenced accordingly, with the result that telephone activities took different forms in relation to the areas in which the system expanded, and the social practices that it helped to reproduce were differentiated by class and gender. In practical terms, the spatial distribution of telephone systems influenced the kinds of uses people made of them.

Since, as pointed out earlier, Bell Telephone Co. monopolized development of telephone systems in profitable areas, its policy was to supply the best technological apparatuses where it was economically and politically advantageous, and cheaper material in less promising locations. The areas to which Bell refused to extend its system were taken over by independent telephone companies, mostly in rural areas, where they thrived from the early 1890s. In 1885, part of Bell's patent was declared void. Although this affected only a part in the receiver and not the most important element of the technology, it left some room for independent companies to expand. When the American patent expired, in 1893, independent companies could buy their equipment from American manufacturers that were not related to Bell. Between 1892 and 1905, eigthy-three independent telephone companies were created in Ontario alone (Babe 1988, 17). After the telephone inquiry in 1905, the Railway Act of 1906 brought state regulation to development of telephone systems, obliging Bell Telephone Co. to connect independent companies' networks to their long-distance lines. Six hundred and seventy-six mainly rural independent telephone companies were established in Ontario between 1906 and 1915 (Babe 1988, 18).[1] These were small firms, often developed spontaneously by people living in the area, which charged very low rates for the service. From this uneven expansion emerged diversified systems. Various types of telephone systems installed in different places reproduced divergent communicational forms. In rural areas, the relationship between the development of a specific telephone network and the form of communication it engendered was particularly distinctive. Several rural telephone companies were co-operatives whose structure was determined by the users themselves. This contrasted with the more rigid system established by Bell in urban areas.

PRIVACY IS LIBERTY

Very early, specific telephone practices influenced people's daily activities. At the same time, a controversial issue was raised by this new set of telephone practices: the capacity of the telephone to reproduce the degree of privacy already extant in the social practices of the ruling classes. The accommodation of telephone systems to private communication, in order to satisfy these classes, certainly affected the types of interaction between users, and differentiated them from those occurring on party lines. Thus, the different kinds of telephone lines provided to subscribers influenced the form of long-distance interactive communication in a community, creating various telephonic practices, some of them unexpected, which led to a "culture of the telephone."

The ruling classes in late-Victorian society had a highly developed sense of privacy. To many members of these classes, the idea of having a conversation overheard by eavesdroppers was in itself a limitation to "freedom of speech." Since the telephone was financially available only to the wealthy classes, especially in urban areas which were mostly controlled by Bell, the nature of its service and the form of communication it created were determined by the social requirements of these classes, notwithstanding the needs of other social groups. Privacy in bourgeois and petty-bourgeois telephone practices was preserved at two levels: secrecy in telephonic communication, and protection of the household from intrusive telephone calls.

During the early period of telephone development, privacy was not particularly problematic, since telephone networks took the form of several "domestic lines" linking small groups of households and businesses, and constituted a supplementary link within an already entrenched social group. They were formed despite the fact that Bell Telephone Co. did not advise "the use of more than two [stations] on the same line where privacy [was] required" (BCA, sf.ind 1877). However, since exchanges were not yet in use, and telephonic communications were possible only between people connected on the same line, private lines were not practical because they limited the possibility of contacts to two places. A telephone system without a public aspect did not have a significant use value. As a result, groups of two or more businessmen connected their offices and households onto one line, which was primarily used for business transactions. Women were not expected to use the precious technology for more than a few minutes a day, to order supplies or organize social engagements. Those who monopolized the line for a chat with a friend or a relative at times when it was required for business were likely to be disciplined. For example, a man who had tried for about ten minutes to get his telephone line without success because his wife was talking to a friend gave her a lecture when he finally reached her, and subsequently had his residential telephone removed (BCA, qa 1919, 112). Another businessman, from Burford, Ontario, also had his home telephone "removed at once" after his wife had called him at work to ask to borrow five dollars! (BCA, d 12113, 1878).

Privacy became more problematic when all of these domestic lines were finally connected through a central exchange which indiscriminately linked different groups of subscribers on party lines. Not only could people no longer choose those with whom they shared their line, but the bad quality of the wires produced cross-talk which allowed subscribers to hear conversations on other lines as well as on their own. As one writer pointed out, these "outside interferences" with private conversations were "annoying" (BCA, d 12016, 1884a, 1). The press began to emphasize "the danger [that] the telephone" would put "an immediate end to all privacy" (BCA, d 12015, 1877c, 12). In fact, some of the complaints were justified, as the "remedies" suggested by telephone companies to counter abuses on party lines were usually not followed. "Listening on the line" was generally practised. The result was that the telephone was seen as an "indiscreet instrument" (BCA, d 12016, 1880d, 25). The solution to the problem of privacy appeared to be found exclusively in private lines, since secrecy devices attached to telephones were not promoted by the companies.[2]

Indeed, some subscribers were requesting private lines. Despite their expense, "the majority of our customers in Montreal are demanding separate lines on our exchange," wrote McFarlane to Swinyard (BCA, d 1173, 1880). High rates did not deter wealthy subscribers; for them, privacy was worth the money. However, a different problem plagued the promoters and consumers of private lines: the mediation of operators. Operators were accused, rightly or wrongly, of all the evils of indiscretion. As one irate subscriber claimed,

Whatever is said in the secrecy of the piazza by youthful students of the satellites of Mars will be proclaimed by the way of the housetop to the eavesdropping telephone operator. No matter to what extent a man will shut his doors and windows, and hermetically seal his key-holes and furnace-registers with towels and blankets, whatever he may say, either to himself or a companion, will be overheard. Absolute silence will be our only safety, conversation will be carried on exclusively in writing, and courtship will be conducted by the use of a system of ingenious symbols. An invention which thus mentally makes silence the sole condition of safety cannot be too severely denounced, and while violence, even in self-defence, is always to be deprecated, there can be but little doubt that the death of the inventors and manufacturers of the telephone would do much toward creating that feeling of confidence which financiers tell us must precede any revival of business. (BCA, d 12015, 1877c)

These attitudes had both a class and a gender context. Most subscribers from the ruling classes considered working-class operators to be "devoted servants of indulgent male overseers ... [and] intruders of dubious ability and fragile reputation" (Marvin 1988, 26). Because these women needed work, they were considered untrustworthy. Moreover, women in general were thought of as being afflicted with a compulsion for listening on the lines which they could not control. These opinions persisted in spite of the denials of telephone companies. Bell Telephone Co. invited journalists to visit the exchanges in order to show them that "the operators ... [had] too much to do to pay attention to the conversations that [were] passing, and thus the people [had] the whole thing entirely to themselves" (BCA, 12016, 1884b). The operators' consistently bad reputation in relation to privacy was detrimental to the company's expansion, as it deterred some people from subscribing.

Just as eavesdropping was eliminated by private lines, the remedy for operators' listening to check the lines was found in automatic telephony. In the early 1880s, a newspaper predicted that the problem for "every man who desires secrecy for his communications" would be solved only when he "will be is own operator" (BCA, d 6453, nd). In 1905, when the first automatic telephones were marketed in Canada,[3] newspapers claimed that the "modern marvel of telephoning" gave a "delightful sense of privacy" (BCA, ncm 1880–1905). Its major advantage was that "it guarantee[d] an absolute secret transmission of all conversation" (BCA, nca 1911a), "as closely guarded as though two persons spoke together in a brick walled room" (BCA, ty 9 (5) 1905, 390).

It did not take long, however, to discover that even automatic telephony could not ensure *perfect* privacy, and that the practice of eavesdropping could not be attributed solely to operators. With private lines and automatic telephones, eavesdropping gave place to wire-tapping by various parties, including telephone companies. Indeed, during the 1907 Ontario Select Committee, Mr Maw of Bell Telephone Co. did not deny the existence of a "listening board" in the company, and said that "no girls refused to do *listening duty* except for the long hours such work entailed" (BCA, ncm 1907b, my emphasis).[4] The *Globe* asserted that "the company ha[d] the machinery for a system of espionage more than Russian in its perfection" (BCA, ncm 1907a). Police forces also revealed that "the telephone service [was] invaluable" in social control, and that they were "entitled to every aid," including wire-tapping, "in frustrating the plot of crooks and confidence men" (BCA, nct 1916e). Given such violations of the "right of a man to privacy," provincial governments decided to act. In 1917 in Ontario, and in 1918 in Quebec, bills were passed to regulate eavesdropping and wire-tapping on the telephone, imposing severe penalties – one hundred dollars or three months' imprisonment in Quebec, twenty-five dollars or thirty days' imprisonment in Ontario – for offenders.[5] (These regulations, however, did not explicitly cover wire-tapping by the police.[6]) This led to a process of legalization of the right to privacy in telephone calls. What started as a *domestic* issue became a provincial – and later a federal – *legal* question. Legislation was passed to regulate what was heretofore ruled by *etiquette*, and *disciplining subscribers changed to controlling users*. Privacy, then, an issue of overriding importance to the ruling classes, became institutionalized as a general feature of the telephone system applying to all classes, notwithstanding their different social practices and cultures.

Although the issue of secrecy over the wires was considered the major problem, there were other forms of violation of privacy attached to telephone use. In the early days, weak transmission obliged people to shout into the instrument in order to be heard at the other end. Public telephones in drugstores provided no possibility of privacy for the working classes who used them. Some businessmen and professionals refrained from using the telephone because it did not allow for quiet conversation – "A telephone caller had to shout as if he were speaking to another person 80 feet away" (BCA, bb 1951).

The development of the telephone also created encroachments on privacy at another level. People thought that the telephone was a terrible invader of domestic intimacy. Once they were connected with an exchange, the question was "how far each householder [would] be *at liberty to reject the temporary union*" (Bedford 1879, 413, my emphasis). Housekeepers and their families "complain[ed] that when they [were] busy they [were] continually being rung up about trivial matters" (Hastie 1898, 894). As one writer pointed out, "The doors may be barred and a rejected suitor kept out, but how is the telephone to be guarded?" (BCA, ty 10 (3) 1905, 221)[7]

The fact was that late-Victorian women were caught off-guard. The barriers that their society had built in order to preserve privacy did not work with the telephone, and there was no time to construct new ones. Yet, in spite of this inconvenience, women continued to use the telephone, and the system developed rapidly, especially from the early 1900s onward.

CREATING "STANDARD" TELEPHONE PRACTICES

Telephone practices related to recreational uses were introduced slowly as the telephone developed, springing from the form of communication created by telephone systems shaped by men from the ruling classes. These men prescribed women's early recreational uses of the telephone. As Marvin says, "Male control of female communication was justified by women's ignorance and should have guaranteed it as well" (1988, 25). The recreational telephone uses specified by male managers were to be *rational* activities – "appropriate" uses governed by an ensemble of rules and procedures. Female consumers did not necessarily agree.

Daytime telephone service appeared in cities and towns toward the end of the 1890s, with the extended use of copper wires, and night and Sunday service was provided to all exchanges with more than one hundred subscribers (BCA, bb 1950). Although night service raised relatively few objections, Sunday service caused much controversy in Toronto. The president of the Toronto Ministerial Association wrote to "the president and directors of the Bell Telephone Co." in 1881 to urge management to keep the exchanges closed on the Lord's Day, arguing that "very much of it [the Sunday telephone business] is not justified by any requirement either of necessity or mercy," and therefore "the Association desires very respectfully to urge your board to cease keeping the office open for business on the Lord's Day" (BCA, d 9498, 1881). The Association was supported by the Toronto *Globe*, which deplored Bell's encroachment on the day of rest. According to the newspaper, there was "no sufficient reason for depriving the young women employed in the office of their weekly rest." Since all places of business were closed, the *Globe* found there was "small occasion" to use the telephone on that day; its most common use was for hiring a cab, which, according to the *Globe*, could be done the day before (BCA, nct 1881). In spite of such public opposition, however, the telephone was very much used on Sundays as well as on weekdays; at the same time as operators' working hours were extended to Sundays, an intensive advertising campaign suggesting other uses for the telephone began.

Very specific telephone uses were prescribed by the companies: shopping, making appointments, protection, and personal conversations. Each recommended telephone activity was confined to a particular period of the day. During the day time, when the lines were "indispensable" for business, housekeepers were requested to restrict the use of their telephones to shopping and to short calls to arrange social engagements. In the evenings, when business traffic was so low that telephone companies offered special rates to encourage consumption, women were permitted to call friends "for a chat." Finally, the threats and mysteries of the night could be kept at bay by the use of the telephone to summon the police, the doctor, the fire department, or other services (BCA, nca 1911b). "The night calls," said a writer in 1914, "are laden with portent" (Husband 1914, 331).

A great number of advertisements were related to the use of the telephone for shopping. This speaks to Strasser's notion that the companies "linked the activities of the consumer housewife to their own through advertising" (1982, 251). Telephone advertisements were oriented to increasing mass production and consumption, if not of the telephone, at least of other products, in order to gain new subscribers. Strasser also observes that, as early as 1891, some advertisers had already decided that "women made the purchasing decisions," and that advertising was to be directed toward them (Strasser 1982, 244). Bell certainly did not shave this opinion. Indeed, its early advertising, which was aimed at males, explained that the advantage of shopping on the telephone was that it "save[d] car fare, shoe leather, your wife's patience" (BCA, sf.ad 1900). This implied that the savings gained in the use of other commodities – including the wife's labour – covered the cost of the telephone. A few years later, the advertisements took a different tack, appealing directly to the housewives: "WOMAN SLAVES! Enough about household duties and cares without being obliged to run down almost daily for supplies. A telephone would save her time and energies and costs but a few cents a day" (BCA, nca 1903–13). Although the first part of the message harangued women directly, when the price of the product was mentioned the husband was addressed once again, and the wife was referred to in the third person. The capitalists who were developing the telephone systems did not consider women to be their direct clients.

These invitations to shop over the telephone were complemented by department-store advertisements encouraging the use of the phone. In effect, for these businesses, telephone shopping represented an increase of sales coupled with a reduction of labour in the form of messengers. Although the stores had to hire telephone operators for their private switchboards, it can be safely assumed that their number was smaller than that of the messengers, since they could take many more orders during an average day.

The telephone directory also provided facilities for shopping over the telephone. Some people called it "the buying guide" (Lyon 1924, 175); its "first function ... was largely one of publicity" (Lyon 1923, 187). Advertising had a concrete effect on the telephone business. The telephone bill of Eaton's department store amounted to over one thousand dollars per annum in 1899, in contrast to the regular business rate of thirty-five dollars. Sise wrote that Eaton's was not to be affected by the general increase in business rates because the firm was "a large subscriber." Moreover, "there [were] about fifty firms like Eaton, who [were] large users, and [were] not to be disturbed" (BCA, slo.14 (1899): 71). Clearly, by 1899, housekeepers were extensively shopping over the telephone. This activity was so popular that, in the middle of the next decade, some department stores started a service for "all-night orders received by telephone" (BCA, ty 10 (3) 1905, 221b).

The second use of the telephone suggested by Bell Telephone Co. was as a "nightly protection" against unforeseeble situations such as illness, fires, thieves, and so on: "A telephone in your house is always useful, always reliable, and a great comfort. Every housekeeper should have one" (BCA, sf.a 1902); and it should be "on duty day and night" (BCA, sf.a 1904). As early as 1912, the telephone was considered "a necessary part of the doctor's equipment" (*Literary Digest*, 1912, 1037). Even earlier, the police were equipped with "a police patrol system of huts or kiosks with signalling and telephone equipment connected to the police stations" (BCA, hit 1877–1909; tg 2 (2) 1910, 9). The police could make arrests with the aid of a telephone call (BCA, qa 1881a, 48). It was said that the police department saw the telephone as "a very important part of the city's police system" (BCA, ty 10 (4) 1905, 294). In fact, one of the first expected uses of the telephone was as an adjunct to law-enforcement agencies responsible for social control. Only the fire department was not adequately provided with telephones (BCA, nct 1914c); people continued to use fire-box alarms until the 1940s.

An interesting feature of Bell Telephone Co.'s advertising policy was that some telephone practices that had been considered unnecessary, and even unjustifiable, in early periods were later legitimized. Lengthly chats on the telephone, for instance, were strongly condemned prior to 1890, and disapproved of before 1900. Advertising suggesting use of the telephone for a chat began with the expansion of private lines. Although Bell never explicitly stated that chatting over the telephone was limited to private lines, the rules for "etiquette for party lines" recommended that the latter be used only for indispensable, short calls so that the lines would be available for more serious purposes. The fact was that what was called "gossip" by male journalists and company managers – but what really consisted of conversation between friends – was still "a subject of jesting and scorn." Nonetheless, the alternative, the "numbing solitude of hours of loneliness," was not considered "elevating and edifying" either (BCA, ncm 1906c). In any case, advertisements promoting the use of the telephone for a chat were rare. The first one appeared in 1911, followed, in 1912, by another one presenting the telephone as "a very comforting thing to call friends and relatives ... and have a fine chat." The accompanying illustration portrayed an evening scene (BCA, nca 1912, 85).[8]

Later, inspired by women's recurrent use of the telephone for sociability, the telephone companies modified the discourse in their advertising, and presented the telephone as a psychological support against loneliness, stress, and fatigue. An extension phone, for example, saved labour, time, and thus, nervous strain (BCA, nca 1911a, 47). Ads presenting the telephone as a psychological aid suggests that telephonic communication had become a "way of living." No longer was the telephone a mere accessory to daily physical domestic chores; it was becoming an integral part of the housewife's life, transcending it, and regulating her psychological activities, her unconscious. In M. Mattelart's words, it rendered "her exile more gentle." Since the advertising was addressed primarily to the husband, he became the conscience of the household, purchasing a telephone to give his wife more rest and his family more happiness. The telephone had become a "living thing with creative and transformative powers." This perfectly fit a McLuhanite scenario in which the medium was the message: the technology unilaterally transformed society. The role of the ruling classes in determining the pattern of distribution of telephone systems and in controlling the production of telephone calls was completely hidden by the implication that the technology itself exclusively had that power.

These advertised practices were generally approved by the social groups in which they developed. The prescription of *standard* uses in advertising implied that other ways of employing the telephone were not acceptable to those who controlled the systems. However, the restrictions were not always respected, and some subscribers used their telephones for "unreasonable" activities. A large number of these users were women who did not accept the telephone company's prescriptions. In fact, these "delinquent" telephone activities created by women were largely responsible for the change in the company's advertising policies over the years.

By the early 1900s, women from the bourgeois and petty-bourgeois classes were using the telephone extensively, not only for shopping or other indispensable activities, but also for social purposes. A detailed agenda of telephone uses by women during that period[9] shows that women's use of the telephone for a chat was extended over the entire day (BCA, ncm 1907e). This type of use became part of women's social practices, and had some influence on the development of the telephone, not only in terms of the code of telephone practices but also in terms of the pattern of distribution of the systems. Indeed, while early development was planned exclusively for business areas in cities and towns, women's use of the telephone soon obliged Bell to revise its plan and to take domestic development into account. Urban sectors that hitherto had been overlooked began to look attractive to the company. Later, houses were equipped with extensions or supplementary lines in order to allow for the husband's business calls as well as the wife's social calls. Most of these changes were due to unexpected practices.

UNEXPECTED TELEPHONE PRACTICES

Among the new social activities developed by the telephone and practised by women, phoning friends and relatives was certainly one of the most popular. Conversing over the telephone was seen as "taking the place of visiting" (Spofford 1909). It was faster and more convenient than having to harness the horses and, sometimes, convincing the husband to make the journey. Although it is impossible to determine the percentage of residential calls made just to chat, complaints published in newspapers and magazines about women's habit of talking on the phone for "futile motives" (BCA, ncm 1919g) suggest that the telephone was regularly used for that purpose.[10] Motives for making calls included chatting, courting, discussing, gossiping, and so on. This activity came to be so popular that some newspapersmen called the telephone "our tap of communication" ("Back to the Land" 1906, 530). Since, in spite of telephone-company advertising, these calls were made at any time of the day, they multiplied contacts with friends or relatives, the more so since they did not require any preliminary preparation such as change of clothes. "Telephone service enables morning gossiping ... afternoon visits to be paid without the necessity of dressing up or of driving on a dusty road in the hot glare of the summer's sun, or in the biting winds of a wintry day; evening visits to be returned while reclining in one's own comfortable rocking chair" (BCA, ty 9 (3) 1905, 257).

This was a significant improvement for women of the 1890s, since getting dressed was an elaborate and time-consuming process for them. According to Haller and Haller, "it was the duty of every [middle-class] woman to look as beautiful as she possibly could" (1974, 141). For the Victorian middle-class woman, "cleanliness was next to Godliness," and she was "continually advised to keep herself spotless" (Haller and Haller 1974, 145). As a consequence, she "redressed several times during the day," each time tightly bound in corset, bustle, petticoat, and extravagant dresses, in order to receive visitors, to go out to visit, or simply to "await [her] husband['s] return" (Haller and Haller 1974, 161). Telephone visiting diminished the number of "visual" contacts necessitating a change of clothes. At the same time, it permitted these women to remain in "talking" contact with each other.

The possibility of several telephonic contacts per day was said to put women "on the tenderhooks of expectation and desire": the expectation of being "called up" by someone, and the desire to call someone else up. "Thus may life be made miserable by the very attempts to make it easy and happy," said a male writer in *Chambers's Journal* ("The Telephone" 1899, 313). Use of the telephone, like that of the bicycle, was seen as a moral issue necessitating a specific set of rules. Indeed, both technologies became popular with women in the 1890s. The bicycle was considered a "curse" because, like the telephone, it provided women with "evil associations and opportunities" for contacts with strangers without the presence of a chaperon. The use of both technologies by Victorian women, then, had to be controlled by "correct etiquette" elaborated by men, who considered that, "in their weakness," women were "best protected in the privacy of the home."[11] The etiquette was intended to prevent women from using these technologies for "undesirable" and "dangerous" practices.

The "social" aspect of telephone technology had not been foreseen by the early capitalist developers of the telephone system. It is legitimate to assert that the popularity of the telephone with women was partly due to several technological characteristics specific to this means of communication. For instance, the sense of privacy created by conversations transmitted from ear to ear and involving the whole person, to borrow McLuhan's words (1964, 240), endowed telephonic communication with a kind of intimacy which women had not previously experienced. Since, in addition, telephone service was developing into a private-line system in large cities, a conversation on these lines took the form of sharing a secret. However, on party lines, which were the majority in small towns and villages and still numerous in cities, women had quite different telephonic experiences, and were attracted by other features of the means of communication.

In rural areas, the independent telephone companies that developed party lines applied much looser rules to the use of their telephones and charged much lower rates, so that in many areas almost everyone could afford a telephone. Moreover, rural communities were more closely knit socially than urban ones, although they were more sparsely distributed geographically. A letter from K.J. Dunstan, local manager in Toronto, discussing the possibility of opening an exchange in the Beaches area (which was still a rural district at the time), asserted that there was "considerable local intercourse" between the inhabitants (BCA, sb 84141b, 3146–3, 1902).[12]

All of these elements helped generate different types of telephone activities. What was considered rude and "unethical" in the set of rules specifying approved uses of the telephone became helpful behaviour within the code of unexpected practices. These represented a complete reversal of the standard uses – so much so that big-company managers were scandalized by the practices allowed on rural party lines, saying that "no company which ha[d] the best interests of itself and its subscribers at heart, [would] operate them," because they did "not embrace the highest ideals of telephony." On the other hand, some small-company managers thought that "the party line was a necessity and ha[d] come to stay" (BCA, ty 7 (6) 1904, 453). Some users eavesdropped and participated in other subscribers' conversations. The operator of the exchange of the small telephone company owned by Dr Beatty recounted that he "liked to listen in on the conversations ... and would often feel moved to break in and give his views on the topic under discussion. This would have disconcerted town or city folks, but the doctor's subscribers ... knew his ways and took this in their stride" (BCA, d 29909, 1967).

Actually, in the code of rural party-line activities, listening to others' conversations was not seen as eavesdropping by subscribers, but rather as participation in community life: "Every country user did [it] ... it was the way they got the news" (BCA, d 29909, 1967). Often, in small communities, a listener entered a conversation with information which the two original callers did not have. For instance, *Telephony* reported that when a woman cut her finger while cooking dinner and phoned a friend to ask for advice, "before the friend could answer someone else piped up, 'Bind it up in salt pork.' Still another voice advised court plaster and someone else had another remedy to offer" (BCA, ty 8 (3) 1904, 211). Most of the time, though, listeners tried to go unnoticed, just as they would if they were eavesdropping on a conversation in a public place. When a man called a friend to announce his visit, he added at the end of the call, "'The rest of you on the line – Martha, Grace, Mary, Rachel – tell the men I'll buzz wood tomorrow afternoon.' The men all appeared and there was no explanation asked or offered about how they knew when to come" (BCA, qa 1936). Although this example implies a sexist tendency by presuming that women, and not men, were the listeners, it shows how party lines were used in rural communities. As one observer pointed out, "The strange part about a party line in the country is the fact that everybody listens but very, very few ever admit that they do" (BCA, qa 1936, 51). People knew that they were often overheard, but most of them did not mind. They knew that, in time, *they* would be the listeners. It was part of rural life.

One of the most important characteristics of party lines, especially in rural areas, was that they were regularly used for "meeting on the lines." For instance, when eavesdroppers decided to enter a conversation initiated by two other parties, the telephone call generated a group discussion: "It is ... evident ... that if one person calls up another in the far end of the town many receivers between these two points come down and sometimes more than two persons join in the conversation," the manager of an American independent company remarked (BCA, ty 8 (3) 1904, 211). Sometimes, the technological features of the telephone network were responsible for these meetings. Indeed, some small companies did not have a discriminating ringing system – the same ring applied to every house – so that when the telephone rang, all subscribers had to answer to check if the call was for them. Often, several users stayed on the line to participate in the conversation (BCA, qa 1918, 120; d 29909, 1967; d 29912, 1961). At other times, the operator was asked to connect a subscriber with several others, instigating a meeting. One operator recalled that she "would connect two or three lines and hold them open so the women could talk back and forth and arrange church meetings or other projects" (BCA, d 29909, 1967). Sometimes, a woman would keep the telephone receiver to her ear while she was working: "There sat his wife in the rocking chair. She was sewing and tied to the back of the chair was the receiver of the telephone, so adjusted that she could place her ear to it without changing her position ... it enabled her to hear the gossip of her neighbors at the other end" (BCA, ty 6 (6)

1903, 480). Finally, party lines were also used to comfort the sick. The telephone receiver was placed on the ill person's pillow so that he or she could listen to conversations on the line and keep in contact with what was going on in the vicinity (Spofford 1909). These examples show some women's initiatives to decrease the loneliness they felt in their isolated homes. For them, the telephone was a means of staying in touch with the rest of the community. They did not need to participate directly in all activities occurring over the phone. In fact, before the advent of the radio, the telephone was the only way for these women to hear other people's voices without having to leave their homes. Most men, however, ridiculed, or altogether dismissed, these ways of using the telephone to improve women's lives.

The unexpected uses of the telephone practised by women influenced the companies' notion of its value. This technology, which had been conceived exclusively for business, seemed to have alternative uses that were worth considering. However, among these uses, only those approved by management were retained. For instance, collective calls, regularly practised by women on party lines, were gradually replaced by private lines and telephone calls between two parties.[13] However, of the practices retained by the companies, some had been created by women. One of them was the use of the telephone for sociability.

This suggests that if women had restricted their use of the telephone to that promoted by the companies, today it probably would not be so an inconspicuous technology in the household. Indeed, at the domestic level, it would still be a form of communication to be used on special occasions only. Yet, although the telephone system was adjusted to take into account some activities practised by women, it was not planned primarily for them, as their social and cultural practices were not directly taken into consideration in its expansion. Here, it is useful to use Cockburn's concept of "male tenure" over technology to explain the participation of women in the structuring of the telephone system. In her article entitled "The Relations to Technology" (1986), Cockburn suggest that men have what she calls a "tenure" over the technological sphere, which means that they "appropriate and sequester" each new area of development at the expense of women. This appropriation by men is manifested not only in development and ownership of technology but also in its uses and values, which are, according to Cockburn, mostly determined by men. She argues that "technological competence correlates strongly with masculinity and incompetence with feminity" (1986, 78). In the telephone values developed by dominant-class males, women's specific uses of a telephone system developed by and for men were clearly deemed incompetent. Women's persistence in using the system their way, and the lure of profit that these unexpected female practices represented to the telephone business, finally resulted in the development of a service that was better adapted to women. Thus, as users, women had only an indirect impact in the pattern of development of the telephone. However, their contribution was an *active* one, since some of their telephone practices forced the companies to modify their development strategy. In addition, the various uses made of the telephone engendered some social change, and a culture of the telephone was slowly developing.

THE TELEPHONE CULTURE

The elaborate system of telegraphy that existed before the advent of the telephone served those who later became telephone users. The telegraph, which constituted an important improvement in terms of speed over letter-writing, had been used extensively for almost fifty years. When the telephone began to be marketed, however, the telegraph came to be seen as a slow means of communication. Transactions which took days to be made by post, and hours by telegraph, could be completed instantaneously by telephone. Telephone companies' advertisements stressed the speed of the telephone in comparison to other means of communication. "The mail is quick, the telegraph is quicker, but the long-distance telephone is instantaneous and you dont [sic] have to wait for an answer," said one (BCA, d 1544, 1898). These particularities of the telephone influenced social practices. Although, as some claimed, the telephone had not "revolutionized the modes of correspondence" (BCA, d 12016, 1879b, 10), it did modify several cultural practices.

The telephone did not supplant existing means of communication. As a writer pointed out, "A letter was different from a conversation ... In a letter, you could get down on paper exactly what you wanted to say in the best possible language, and leave out whatever didn't fit it. It was like addressing a jury without the presence of opposing counsel, in some courtroom where you had a free hand with the judge" (Langton 1987, 82). Whether for this reason or because it was less expensive, written correspondence was still extensively used. In 1905, for example, while the telephone had superseded the telegraph for short-distance communications (e.g., communications within a city), the latter was still generally used for long-distance transactions (BCA, ncm 1905a). The postal service was also regularly utilized. The rush that Bell Telephone Co. experienced in 1918, during a postal strike (BCA, nct 1918b), was evidence of massive use of the mail system at the time. Yet use of the telephone was growing rapidly all over the world (see figures 10 and 11).[14] It had evolved from being seen as a "nuisance" and an "indignity" to being a "sign of civilisation." "Failure to adopt the use of tele-

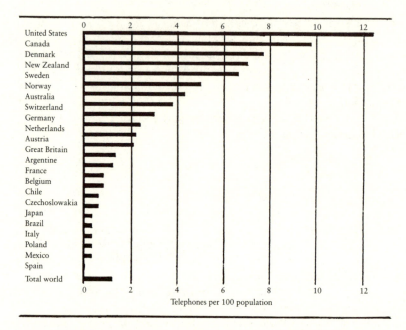

Figure 10: Telephones per 100 Persons in Selected Countries, January 1921.
(*Bell Telephone Quarterly* 1922, 1(3): 49)

phones," said a writer in 1905, "indicates, in general way, a backward condition, a lack of enterprise, in any modern city" (BCA, ty 7 (6) 1904, 456).

Extensive utilization of the telephone by the wealthy classes was bound to create some specific habits. Actually, use of the phone was affected by the time of the day and the weather: "The more inclement the weather, the more of people resort to their telephones. There are appointments to be cancelled or deferred and taxicabs to be summoned" (Rhodes 1929, 21). Some women would rather phone their friends than go out in the rain or the snow to visit: "We can tell what kind of weather it is from the College Exchange," said one operator, referring to residential calls (BCA, nct 1914b).[15] An observant operator divided the daily activities of "leisure class" women over the telephone as follows.

At seven o'clock, there are scattered calls ... for doctors ... At eight o'clock, the nice, early-morning women come on the market with patient, affable butchers ...

At ten, interminable communications between women ... with infinite details as the clothes ... I've known them to keep it up for three quarters of an hour.

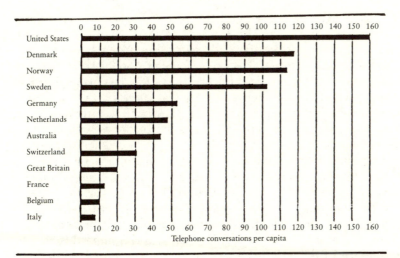

Figure 11: Telephone Conversations per Capita in Selected Countries, December 1920. (*Bell Telephone Quarterly* 1922, 1(3): 50)

At eleven to half past ... nippy ladies calling up employment agencies, or stupid servant girls replying. At eleven thirty till twelve thirty there's a wild rush, everybody trying to catch everybody else for lunch.

From then till three or so there are characteristic calls of all sorts: peevish, hurried females who use the nickel 'phones in downtown drug stores ... silly school girls mischievously calling men they don't know ...

From three to four ... a flurry of women trying to call up stores before they close, or in the catch of the last deliveries.

At five, wives begin to call up to know if husbands are coming home ... 'Be sure to bring home a steak or a lobster.'

From six to seven everybody seems to be busy to call up ... a club ... a garage ... towards eight, comes the nervous maiden[16] calling up her men ...

After ten thirty come the carriage calls, garage orders, and the hotel private exchanges begin to get busy. (BCA, ncm 1907e)

Women used the telephone for various purposes. It was said that a woman "no more need[ed] to make appointments by letter with the dressmaker, or to drive to the box-office of the theater to take tickets, or to be kept waiting for forty-eight hours before she knows whether Mrs Blank can meet her or come for tea" ("Back to the Land" 1906, 530). It was faster to use the telephone, and get what she wanted without leaving her home. The fact that women were using the telephone in this manner meant that they no longer expected to meet a regular group at such locations as the market. The telephone was taking the place of the daily shopping trip, at least for some women, especially on inclement days. Housekeepers were slowly changing their daily habits, thereby modifying the characteristics of the places they used to patronize.

Technological features of telephone systems also contributed to the development of certain cultural practices. The fact that the phone allowed oral communication without visual contact created a kind of intimacy which people previously had not experienced (Barrett 1940, 129). However, these features had some drawbacks as well. Having a conversation ear to ear did not always create the desired intimacy. It was reported in some scientific journals that this mode of communication sometimes generated insecurity, especially when the person calling was unknown ("Action at a Distance" 1914, 39). Indeed, it seemed that the anonymity provided by the telephone had evil effects on the personality of some callers, so much so that the telephone was seen as having "a brutalising influence": "The sensitive-minded man who would shrink from saying a disagreeable thing in ordinary conversation, when talking through the telephone, will speak his mind ... bluntly and argue ... roughly ..." (BCA, ncm 1906d). Thus, the telephone was said to encourage the use of foul language (BCA, nct 1916d) and "trespassing" by telephone. Some men importuned women over the telephone in such a persistent manner that judges deemed the offence a "breach of the peace" (BCA, ty 10 (3) 1905, 221). Thus contradictory effect created by the telephone of feeling nearby and far away at the same time seemed to embolden some people, leading to new breaches of the law to which the legal system had to adjust.

In fact, the problem of aggressiveness and "foul language" on the telephone became so serious that an amendment to the Telephone Act was passed in 1915 naming the use of "abusive language on the telephone as an offence punishable with a fine of $25 or imprisonment for 30 days" (BCA, nct 1915a).[17] Whether the telephone was entirely responsible for such behaviour is debatable. Industrialization of society was causing rapid changes in some social practices and encouraging more liberal and emancipated social behaviours. Since the instantaneous characteristic of the telephone constituted "a means of projecting personality," as stated in an advertisement (BCA, nca 1925), without the necessity of identifying oneself, it permitted some hidden features of a personality to surface. "The use of the telephone gives little room for reflection," stated a writer in the *Chambers's Journal* in 1899. "It does not improve the temper, and it engenders a feverishness in the ordinary concerns of life which does not make for domestic happiness and comfort." The telephone, by making life "so easy," represented an "immanent danger of relapsing into barbarism" ("The Telephone" 1899, 313). This notion imparted to the telephone a responsibility which should have been attributed to the social conditions created by industrial capitalism as a whole. The *only* contribution of the telephone was to *facilitate*, through the anonymity it afforded, the emergence of unpleasant characteristics which already existed.

Such evils led to the necessity to develop *telephone etiquette*. Telephone etiquette was elaborated from the standard uses prescribed by the telephone companies. People were told to use good manners on the telephone, to employ such general courteous phrases as "Please" and "Thank you," to apologize for making callers wait, to utilize "correct" language instead of familiar expressions or abridged sentences (BCA, ty 8 (2) 1904, 130). Users were also advised to answer their telephone themselves to avoid making the other party wait for them (BCA, ty 8 (4) 1904, 311), and to identify themselves when answering or calling (BCA, ty 8 (2) 1904, 130). Failure to follow telephone etiquette was seen as a matter of gender, as women were presented as the main offenders in terms of telephone manners. Operators reported that women callers "have an exasperating way of asking, 'Who is this?' when some one answers their call whose voice they do not recognize." "Girls" were accused of unduly using their employers' telephones during business hours (BCA, ncm 1915a). The ultimate abuse, however, was attributed to women who did not have a telephone at home and who used telephones in drugstores. These women were seen as the "chief patrons" in "carry[ing] on all the conversation you wish" of "the most trivial nature." They went to the nearest store and, with a "May I use a telephone, please?," they used the storekeeper's telephone and "for 20 minutes or half an hour they will carry on the most milk-and-water-conversation" (BCA, d 30114, 1965). The ultimate offence was that they left the store "without spending a cent."

Still, even when "good manners were observed," telephone calls were considered "hopelessly vulgar" for "ladies of the high society." In her book *Etiquette for Americans* (BCA, qa 1906), a "Lady of Fashion" claimed that the telephone call, due to its instantaneous character, was a "blessing in adjusting details" for a reception. However, it "should be used sparingly." Informal "invitations to bicycle or play golf [could] be transmitted in this way ... but for most social matters, the use of the telephone [was] questionable." Moreover, there was "no excuse for telephoning an invitation when time [was] not an object, or when the person invited [was] not an intimate friend" (BCA, qa, 1906, 37). This book was written exactly thirty years after the telephone was first marketed. Old cultural practices die hard.

The only concessions made by the "Lady of Fashion" to telephone use were for intimate relationships and casual encounters. In spite of her recommendations, though, the telephone was sometimes used at first for invitations. In such cases, other means of communication were usually employed to confirm the telephone call. In Wharton's *The House of Mirth* (1905), the petty-bourgeois heroine, Miss Lily Bart, used three different means of communication for a single invitation. She first made the invitation over the telephone, the call being answered by a maid. A note written by Miss Bart and delivered by a servant was then used to confirm the telephone call. Finally, she sent a telegram to finalize the whole process. In rural areas, however, etiquette was not as binding, and telephone advertisements suggested the use of the instrument to send invitations for an "impromptu party." Instead of spending an afternoon driving from house to house to invite people, "in less than half an hour, you could ring up your friends, living miles away, and invite them to come, without trouble or fatigue" (BCA, d 21203–2, 1908). There were clear differences in accepted telephone practices between rural and urban areas.

The telephone had other cultural effects, especially in relation to letter-writing. Although people recognized the importance of letter-writing for serious matters,[18] as early as 1906 "the idea of writing a series of letters with a pen and ink, directing, sealing, and stamping the envelopes, and then waiting till the day after to-morrow for an answer simply paralyse[d]" many people ("Back to the Land" 1906, 530–1). Some writers (e.g., Lang 1906) were alarmed by the decreasing popularity of letter-writing. It "seems to be in decay," said Lang, "and no wonder, for few people have time to read a long letter ... Indeed, talk is mainly done through the telephone ... the art of spelling, even, may come to be lost" (1906, 508). Even the government replaced some written documents by use of the telephone. "Government by telephone!" exclaimed Casson. "This is the new idea ... arrived at in the more efficient departments of the Federal" (1911, 899). Lang lamented that it was the end of "the excitement of reading for material in the archives." Since people telephoned instead of writing letters, it would become impossible to trace the development of political, economic, and social organizations, he said. An ex-Chancellor of the English Exchequer confessed that, during his entire career, he had not kept more than twelve letters, most of his business having been done by telephone. Lang exclaimed, "Let us rejoice that the thing was not discovered sooner! If Horace Walpole could have chatted with Horace Mason, in Florence, by telephone; or Madame de Sévigny with her daughter; or Thackeray with Mrs Brookfield; or Mr Stevenson, from Samoa, with Mr Gosse and others, our literature would be poorer" (1906, 507–8).

Undeniably, the telephone has enlarged the field of oral culture. The diary disappeared from most women's lives long ago, and communications between friends occur mostly through the telephone, at least for short-distance interactions. As a result, writing a biography of a person whose life extends past invention of the telephone with the aid of written records only is almost impossible. While the telephone is a technology of rapid and easy contact, it is also a source of transient evidence. For example, it is impossible to know exactly the number and the context of the "visits" paid by telephone during a period of its development. The only sources of information are indirect ones, such as newspaper reports, journal articles, and operators' stories, which may be biased. This means that telephone technology has hampered feminist researchers, for instance, in retracing long-distance friendships between women, which was relatively easy during the time of letter-writing. It is also impossible to trace telephone practices related to working classes. I mentioned earlier that some low-waged women working as maids used, furtively it seems, their employers' telephones. It is almost unthinkable that other members of the working classes did not use public telephones at all, in spite of their poverty. The proliferation of public telephones supports this assumption. However, since they did not have phones in their households, it is difficult to know the volume of use. The extension of oral culture due to the telephone certainly represents an inconvenience for researchers. In fact, Lang suggested that each telephone be attached to a recorder so that future generations could keep track of their ancestors! The records, he said, could be likened to letters.

Nonetheless, if there has been a loss of literature with the use of the telephone, there has also been some gain. Very early in its development, the technology was a source of stimulation for artists, and of entertainment for people. Novels were written in which the plot was based on the use of the telephone (e.g., Sayers 1921). The telephone inspired poets,[19] cartoonists (BCA, nct 1918a), and playwrights. In 1880, for example, George Bernard Shaw wrote a sketch on telephone conversations which suggested that the phone "was a tool for female's gossip" (Brooks 1977, 210).[20] In 1923, literary works on the telephone presented its use as an "expression of female desperation" (Brooks 1977, 218). The constant was that women were rarely presented in a positive light in works about the telephone.

Other effects of the telephone on popular culture were related to its physical features. Indeed, people talked about "the telephone voice" and "the telephone ear." The telephone voice was said to be revealing of "whether gentility [was] a thin veneer or a solid substance," the "thin veneer" being proved when an unpleasant answering voice changed suddenly into "amazing mellowness" upon learning who is calling. Using the right voice over the telephone was considered a "difficult art," because the instrument deprived the voice of its nuances. As a result, only two categories of users could be identified on the basis of their voice: those who rarely used the telephone and whose timidity regarding the technology was translated into a "solemnity of the performance"; and regular users, who were relaxed and talked as if they were having an intimate conversation (BCA, tg 3 (1) 1911, 5). Several people thought that the telephone developed "a soft voice," a well-modulated, "lady-like voice" (BCA, tg 1 (8) 1909, 10; ty 10 (5) 1905, 360). In any case, the voice was regarded as a key element in use of the telephone, and there was general agreement among specialists that the voice itself was influenced by the new technology. The ear was also said to be affected by the telephone. The fact that the cord of the apparatus was on the left side encouraged users to hold it in the left hand and to put the receiver to their left ear. According to some researchers, this caused telephone users to become "left-eared" (BCA, ty 8 (1), 1904, 74). They discovered that those who frequently used the phone had more sensitive left ears. Left-eared and soft-voiced people were thus deemed to be a product of telephone technology.

The telephone was also said to affect physical and mental health. It was seen as a "germ collector," and doctors "urge[d] that the health department compel the telephone companies to equip their instruments with antiseptic devices which would destroy all germs as they entered the transmitter" (BCA, ncm 1906f). The number of articles written on this issue[21] shows that it was seen as a serious problem starting around 1905. Public telephones were considered unsafe because it was thought that they were packed with diphteria, influenza, and consumption germs (BCA, ncm 1908d). It was suggested that hygienic devices be installed to lessen the risk of infection. This perception vanished as suddenly as it had appeared, without any apparent change in the telephone apparatus.

The telephone was also considered a "nerve-racking" technology because of its capacity to intrude on one's privacy at any time of the day. As one woman attested, "I have been called to the telephone three times this morning by some of my friends who just wanted to visit. Twice the bell woke the baby up and once my blackberry jam burned while I was trying to make an excuse to get away" (BCA, ty 10 (6) 1905, 429). Anxiety was increased by the fact that subscribers were instructed by the companies to answer the phone promptly.[22] Night calls were particularly aggravating, to the point that some physicians refused to have a telephone at their bedside (BCA, d 1009, 1934). It was a fact that the telephone was altering a society previously ruled by rigid, well-determined social practices.

X   The changes in popular practices brought about by the technology were instrumental in the creation of a telephone culture. The new form of communication created by telephone systems reproduced some social activities and modified others. One characteristic of the telephone system planned by and for the ruling classes was its speed. Casson said that the telephone had made life "more tense, alert, vivid" (1910a, 231). *Booklovers' Magazine* claimed that the telephone had "doubled pressure, condensed the world, [made] us all next-door neighbours" ("Behind the Scene at 'Central'" 1903, 390). The effect was multidimensional. The telephone was developed in response to capitalist society's requirement for faster means of communication, and it had indeed accelerated the speed of transactions. Moreover, its capacity for long-distance contact gave people the illusion that it had strengthened the nation's solidarity (Carty 1922b, 9), and eliminated class differentiation (Carty 1926a, 2). In reality, it only permitted entrenched social groups to communicate more often and more rapidly. Telephone contacts between members of the working classes and those of the ruling classes always occurred through an already existing rigid etiquette. Moreover, wealthy women on party lines often complained of the bad manners of low-wage women, and pressed the telephone companies to give them private lines. But what was the deeper impact of the telephone on women?

EMANCIPATING WOMEN?

The telephone had contradictory effects on women: it had some emancipatory influence, yet it often contributed to reproduction, and even reinforcement, of sexist attitudes. Often presented as a liberator for women, it was said that "its power to aid in accomplishment serve[d] to stimulate the wife, the mother, into achievements that [made] life worth living" (BCA, ty 8 (3) 1904, 232). The fantastic capacities of the telephone were to liberate "slave women" from domestic chores, and allow them to be more rested, more sociable, and happier (BCA, nca 1903–1913). Still, when women began to use the instrument for sociability, in order to break out of their isolation in the household, men started to object to this frivolous use of the telephone, and to ridicule them in newspapers by accusing them of having a "gossiping instinct" (BCA, ncm 1906g); in journal articles by calling their practices "irrational use[s]" (BCA, ncm 1907a, ty 10 (3) 1905, 211); and even in books (BCA, sf.ind 1895). The clergy also joined the chauvinist movement. In New York, for instance, the Reverend Parks publicly denounced "women of the leisure class who waste their time in unprofitable chatter over the telephone." They were spending valuable time "in idle talk and in gossip ... in calls that were of no value to any one," instead of busying themselves in cleaning their houses and raising their children (BCA, ncm 1908a). However, Ryan (1983) points out that a significant group of these "women of elite status were involved in voluntary associations performing such activities as care of the poor, self-improvement of young men," and so on. It is therefore reasonable to assume that some, if not most, of their telephone conversations were to discuss issues related to these pursuits. However, telephone use by women produced the same sexist reaction as that toward women's gatherings. For men, who "wanted control of all communication conducted through the technology that belonged to them" (Marvin 1988, 24), women did not meet for important reasons, but merely to gossip. The telephone, which was a technology developed exclusively for business purposes, was losing its seriousness with women's practices.

Yet, at the same time as the telephone helped to reinforce male chauvinism, it also contributed to women's emancipation. An observer in the 1890s asserted that "the telephone permitted girls to be bolder in their approach then [sic] it had to be made face to face" (BCA, sf.ind, nd). The new technology was seen as playing a part "in changing the prudish attitudes" of nineteenth-century women. An elderly woman was "appall[ed] ... to see how they use the telephone nowadays." She was referring to her niece and her male friend talking on the phone while her niece was only partially dressed: "The two of them stood talking to one another just as if they were entirely dressed and had stopped for a little chat on the street!" (BCA, qa 1903, 343) Men, and some women, felt uneasy about this new "breeze of liberation." A male writer reported that "the telephone has been instrumental in bringing the young woman of today to a point where her grandmother wouldn't recognize her; that it is in no little degree responsible for her increasing loose manners and looser habits, any mother who takes the time to realize the situation will doubtless agree" (BCA, sf.ind 1921). Sometimes, this boldness amounted to no more than young women calling male telephone operators during the evening for "flirtatious purposes." Although this is not a serious instance, women did become more outspoken over the years when they talked on the telephone. "The telephone gives the flapper courage – and more it permits a girl to lie in her bed and to talk with a man lying in his bed; it permits her half-clothed, to talk with him a moment after its ring had made him hop nude out of his bathtub. Its delicate suggestiveness is not lost in these instances. The most modest girl in America, the girl who blushes even at a man's allusion to his chillblains, once she gets her nose in a telephone mouthpiece acquires a sudden and surprising self-assurance and aptitude at wheeze" (BCA, sf.ind 1921).

Was it the contradictory feelings of closeness and remoteness, creating a sense of intimacy and safety at the same time, that encouraged women to be more intrepid on the telephone? This "impersonal instrument of personal communication," which enabled women to talk without being seen, was undoubtedly disturbing for some men in this period of female sexual repression, when the "vision" of sexuality was thought most crude, and where its expressions were limited to suggestive gestures (Haller & Haller 1974). It seems that the "delicate suggestiveness" created by the technological particularities of the telephone was well adapted to the prudishness of late-Victorian women, since writers reported that they were not embarrassed by sexual connotations when talking to the other sex over the phone. This "audacity" – which perhaps amounted only to an absence of a non-verbal expression of embarrassment which, of course, could not be seen at the other end of the line[23] – appeared to be due to the telephone, since, at social gatherings, women were said to recover their attitude of submissiveness and prudery. In fact, their audacity was very limited. Some researchers (e.g., Maddox 1977) found that women tended to be passive on the telephone, reproducing their behaviour in society. In other words, they called male friends, but did not initiate a new relationship with a man by telephone, preferring to wait for a call.

Thus, telephone development had a mixed impact on women. On the one hand, it granted them some liberation by decreasing domestic chores and freeing them from some social restrictions in mixed relationships. It also permitted them to "visit" by telephone without having to rely on anyone to harness the horses or drive them. On the other hand, access to the telephone may have reduced the number of visits they made to friends and outings to concerts, which were transmitted over the telephone.

In short, a study of the development of the telephone system in relation to women's social and cultural practices points to a contradiction between the privatization and the socialization of women's communication. The public aspect of the telephone enlarged women's opportunities for socialization by allowing them to have instantaneous contact with a much larger number of people. However, these contacts occurred in the privacy of their home, which may have reduced women's opportunities for socializing outside the household.

In any case, women's contributions to the forms of telephone practices, unforeseen by the male inventors and owners, forced telephone companies to rethink their expansion plans. They definitely influenced use of the telephone, shifting it from a strictly business-oriented one to one oriented toward socialization. In addition, women's use *en masse* of the telephone stimulated expansion of the system, not only in business areas of cities and big towns, but in residential sectors and in rural areas.

Conclusion

This book has been concerned with the social nature of mediated communication under capitalism. This mediated communication, mostly provided by a telephone system developed by private capitalist enterprise, influenced various cultural practices. A 1929 article in the *Bloor Watchman*, entitled "The Telephone," described the impact of the telephone on society: "[T]hrough the present complex telephone system flows the life blood of domestic and business life. In every activity, business and social, it plays a part. No other business enters so intimately and personally into the daily life of the community. It is associated with everything which makes for security, happiness and progress." This depicts the increasing integration of the telephone into daily activities by the late 1920s. People's lives changed because of the invasion of this new communication technology. However, since people's activities are class- and gender-specific, so was the logic of their uses of the telephone. This created conflict between the different social groups affected by the development of the telephone.

This study of technology, culture, and gender has been a study of opposed rationalities of experience. This opposition has presented itself in various ways in the development of the telephone. While political and economic conflicts influenced formation of the system, its patterns of distribution and practices, in turn, were responsible for new political conflicts. In this case, the monopolization of the system by Bell Telephone Co. resulted from conditions of development involving not only state intervention, but also opposition from different social groups. This, in turn, led to a specific pattern of expansion and use. There were many possibilities for technological and geographical development of the telephone. In the early stage of expansion, the technology itself could have been oriented to produce either collective or private telephone service. The systems were given characteristics to meet the interests of those who controlled their expansion. While a number of small systems favoured collective use, Bell's expansion was based on privacy. However, in central Canada, the taking over by Bell of most of these small systems, and the necessity for the others to adjust to the standard of the company that provided them with long-distance lines, led most to adopt the characteristics dictated by Bell. As a result, most of the alternative development possibilities were eliminated. Capitalists retained mainly those uses that were advantageous for the reproduction and accumulation of capital. This meant that privacy was prominent, and collective-communication systems were eliminated where it was possible and profitable to do so.

Still, privacy, which was becoming a generally accepted feature of the telephone systems, could be achieved in different ways. Most of the "secrecy" devices developed in early expansion were controlled by the subscribers, who used them when they felt it necessary. Had they received any attention from Bell management, these devices could have been developed into a technological feature allowing as much privacy as did the private lines, which were entirely controlled by the company. However, Bell's pattern of expansion supported the private-line system. The control of privacy thus shifted from the subscribers to the company. This type of system entailed supplementary fees for collective calls in the form of telephone conferences. This shift represented more limitations on access to the service, not only in terms of cost, but in terms of types of use as well, especially collective uses, which seemed to be favoured by some social groups, for which privacy was not a high priority in a useful telephone system. Communal uses were obviously welcomed in some places, in spite of the contradictions created by the predominance of private interests in telephone development. The groups that felt that their interests were not properly served by the system resisted the pressure toward monopoly by protesting.

Capitalist interests controlling expansion of the telephone system were often in conflict with groups which did not see its development in the same way and which were differently affected by it. One of these groups was women, who were affected in two domains: the household and the work place. The telephone industry transformed the occupation of operator into a female job ghetto very early. Although the wages and working conditions for telephone operators were no lower or poorer than those in most other occupations, being an operator, a mediator in the production of telephone calls, did confer a certain social status. This recognition was the *pleasurable* element of the telephone operator's occupation, which turned her job into a "labour of love." In other words, the pleasure procured by social status made her accept more easily the exploitation to which she was subjected and of which she was critically aware. Thus, there was a certain appeal in the "glamour" of her job, "however bitter-sweet it may be when it goes hand in hand with a social and political awareness," to use M. Mattelart's words (1986, 15). Indeed, the operator's role of mediator had a "bitter-sweet" effect on the labour force.

The development of communication technology by capitalist interests is oriented toward ever-increasing speed of circulatioin of capital. As Harvey stresses, in order to accelerate the exchange of capital, developers may have to replace a technology "well before the value embodied in [it] has been realized through use" (1985, 44). Since the motive behind early telephone development was accommodation of businessmen, it was essential to provide them with a means of communication that would satisfy the exigencies of business transactions. Because there was a limit to the speed with which operators could connect telephone calls, they were displaced by a technology that made faster connections.[1] The operator's mediating work, then, increased the status of her job, but also became part of the dynamic leading to its elimination. As a result, the relationship between operators and the telephone companies also implied the former's resistance to the different forms of power imposed on them, which ultimately forced them into destruction of their own occupation.

Sometimes, operators' opposition was organized and publicized to put more pressure on the company. Most of the time, however, their resistance was part of their daily working activities – individual practices, recognized by the group though not applied as a group, designed to annoy either the company or the subscribers. These forms of resistance show that although these groups of women had little control over their working conditions, they could find efficient ways to oppose some of these conditions, without retaliation by the company. As a group, these women had means of resistance based on unspoken and unwritten rules, which were learned by new operators. These practices created a kind of informal hierarchy among the operators, according to which experienced operators controlled the means by which the workers made both the company and subscribers realize that they were human beings and not merely machines.[2]

These opposed rationalities were not limited to the sphere of production, however. Telephone companies and subscribers had opposing views of the expansion of the telephone system and of its appropriate uses. Consumers often disagreed with the companies' development policies, and forced the state to intervene. More than economic interests were involved here, since the distribution pattern of the telephone systems was linked to specific political-economic conditions and to modifications in social and cultural practices. This relationship was dialectical, as specific cultural and social changes, were, in turn, influenced by the particular form of the telephone system. Other telephone activities might have developed in systems allowing for different uses.

Different pragmatic uses were the products of the class, gender, and location of the users. Dominant- and middle-class female users were oriented toward what became standardized telephone uses, suggested by the telephone companies in accordance with the capitalists' conception of a telephone system, and of its proper and improper uses. Since the telephone was an interactive means of communication developed to create or maintain relationships between people, its impact on women isolated in the household was potentially significant. However, although the telephone could be adjusted to some women's activities, it was not planned primarily for them. Their social and cultural practices were not taken into consideration, and the instrument's beneficial effect on them was more in the sense of making their "exile more gentle," as M. Mattelart would suggest, than to help them organize and liberate themselves.[3] Despite this discrepancy between the telephone companies' and women users' logics of use of the telephone, women's telephonic activities kept increasing over the years. Breaching the apparently immovable male logic of telephone use, women of different social and cultural backgrounds created new uses adapted to their activities, activities which might not always have conformed to male-dominated ideology. This is not to say that the telephone would never have come to be used for these social practices, but women consumers certainly accelerated its development toward this type of use.

This book shows that women are not passive receivers of communication technology, as has often been suggested. As M. Mattelart points out, the fact that women adapt a technology planned and controlled by men to their activities and resist prescribed uses puts into question "the monolithic nature of ideological effect of domination." In a study of "women as consumers," for instance, Mattelart shows that working-class women in Chile did not necessarily read media messages the way senders intended, and that sometimes the way these messages were received "denied their internal logic, leading to a roundabout process of consumption" (1986, 14–15). This reversal of the consumption process has also been noticed in the use of the telephone by women. Although there is no determined content in telephone use, the uses prescribed by male developers sent a clear message to women to use the new technology in the way that was rational to those who controlled the business. Women consumers, however, were quick to resist the limits of these uses. Middle- and dominant-class women who could afford the technology adapted its use to their activities, which were not primarily buying and doing business. Most of them had domestic servants who ran their errands, so their main activity was to visit and meet other women. They adjusted use of the telephone to these activities.

Thus, opposition to approved uses of telephone technology was also gender-oriented. Women's rationality in telephone practices was often questioned by male developers and subscribers, but the telephone did not have a revolutionary effect at the level of gender relations. It tended, rather, to reproduce the patterns of male domination. Despite this, and because of female users' resistance, the telephone introduced new codes of practices, and new sets of activities were developed. Some of these were later prescribed by the telephone industry, as much for economic purposes (they boosted expansion of telephone service) as to answer the pressure of women's extensive use of the telephone. However, female telephone activities had some drawbacks for women subscribers. While they contributed to the increase of opportunities for contact among friends and relatives, they also provided excuses to remain at home. The telephone permitted visits, meetings, and even entertainment without having to set foot out of doors.

From this interaction between different telephonic rationalities developed a telephone culture. Women subscribers were largely responsible for the development of a culture of the telephone, as they instigated its use for purposes of sociability. The socialization of telephone use by these women, in addition to the "sense of security" it provided, turned the telephone from a "nuisance" into the "necessity" for the home.

In earlier stages of the telephone's development, people installed telephones, and particularly in their homes, only when they were reasonably assured that such telephones would be frequently used. Now they install them, even though they may be used relatively infrequently, because they are recognised as necessities. They are installed not alone on the basis of their actual use, but on the basis of their *potential* usefullness; not on the basis of the number of calls that will actually be made over them, but on the basis of the *value of a single call*, made at a time when swift and dependable communication is absolutely essential. (Barrett 1940, 133, my emphasis)

This shift from a quantitative to a qualitative use value of the telephone shows the transformation of people's mentality in relation to its use. This new mentality expressed the existence of a culture in which people saw the telephone as a household technology. The various possibilities for telephone use were instrumental in this change in perception. In a telephone culture, use for emergency is not the only, or even the most important, aspect of the service. Most people, particularly women, have a telephone in order to break out of their isolation and to have social contacts with other people without having to leave their home (Singer 1981). Late-Victorian women users helped to create this "sociability" aspect of telephone use.

Thus, women's contribution to the development of the telephone as a means of communication occurred in different forms, according to the nature of their participation. On the one hand, working-class operators were instrumental to how the labour process in the telephone industry was structured. As such, they influenced the form of communication generated by the telephone service. On the other hand, dominant- and middle-class women's telephone practices put pressure on the telephone companies to extend their systems beyond the business world and to adapt them to women's activities. This impact, however, was interactive: the telephone systems developed by the companies modified some of these women's cultural and social practices.

This book, then, has shown the link between the development of a new technology and the political-economic and cultural-ideological conditions within which it grew. I have stressed the importance of applying feminist and political-economic analyses in order to understand the active participation of women in society. The economic incentives and political interventions orienting the development of a technology have an important impact on the pattern of its distribution and uses. In a capitalist society, profitability is the goal behind development of technology by private interests, and state intervention usually sustains the direction of development suggested by the industry. This lack of significant intervention by the state is further illustrated by Armstrong and Nelles' (1986) study of the development of Canadian utilities. Their analysis of telephone systems in Canada shows that the development of systems by capitalist state agencies in some provinces did not take on a different pattern of distribution and uses from those developed by private capitalist interests in other provinces.

The use of a feminist perspective shows that women may contribute to the distribution of a technology despite their underrepresentation as direct contributors to its development. Yet, although the notion that women have no influence on the development of technology is fallacious, their absence from the professions that control technological development (e.g., engineering), particularly at the higher levels of decision making, is a reality that remains to be addressed. Further, women's exploitation in the labour process of the communications industry is real, despite their resistance and opposition. Their participation in that industry is still mostly in low-paid, low-power occupations. Women have yet to make their presence felt in the development of new technologies.

In short, this book shows that technology influences the cultural practices of a society, but that this influence is mediated at different levels.

The first level is the economic and political forces which control its production – namely, its design, its pattern of distribution, and its possibilities of use. Technology is a social product which comprises the ideological and social values of the people involved in its creation and its expansion. As such, access to it may be affected by the practices and social values of the social group that controls its development. The social distribution and uses of technology are class and gender oriented.[4] Different classes may have different access to a technology. Moreover, members of a single class, with equal access, may encounter different limitations in the use of a technology according to their gender. A technology conceived by and for men will not necessarily suit women's activities. It is essential, then, that women contribute to the development and production of a technology which they use extensively.

An analysis of the political and economic motives behind development of a technology reveals specific features that can be attributed to the industry engendering it, while a feminist viewpoint stresses the particularities brought about by women in a specific system. Looking at the dialectical relationship between these two aspects of analysis helps to understand the phenomenon as a whole. It avoids deterministic conclusions in which cultural and social changes are unilaterally attributed to the technology without consideration of the control exerted by developers or the impact exercised by the users.

Notes

INTRODUCTION

1 A short section of Marvin's *When Old Technologies were New* (1988) looks at women's early use of the telephone; Bernard's *Long Distance Feeling* (1982) is concerned in part with the development of the operators' union and, as such, gives a short account of women's telephone practices during the first fifty years of telephone development.

2 For instance, a television set cannot be used as a sewing machine.

3 Silence was gradually imposed on women as operators, however. For more information on this and related issues, see Martin (1989).

4 For more information on the means of communication as means of circulation of capital, see Martin (in press).

CHAPTER ONE

1 Bell Canada's translation.

2 In contrast, the telegraph can be classified as an alternative means of communication, since its use involves the mastery of definite skills, access to which is controlled by a limited group of people. Patten wrote, "The prominent advantage of the telephone is that anyone can use it, since it requires no special skill on the part of the operator; hence telephonic communication may be established when telegraphic communication would be impracticable" (1926, 38).

3 As Mr Radcliffe, of the Bell Canada Historical Equipment Collection in Montreal, pointed out, "The history of the technical development of the telephone in Canada is essentially the history of AT & T in the States." (Interview given by Mr John Radcliffe to the author.)

4 J.C. Watson assisted A.G. Bell in his research on telegraphy and telephony.

5 The bells had also "their importance as a means of revenue," said a foreman in Montreal. Not only did they bring substantial income to the manufacturers, but they benefitted the service by reducing the average duration of calls and rendering the system more productive (BCA, tg 1 (8) 1909, 11). The indiscriminate ringing system tended to encourage "meetings on the line" which often lasted up to half an hour. This issue will be discussed at greater length in chapter 6.

6 Subscribers were accused by telephone companies of abusing the telephone lines with long calls. This device was said to make them more aware of the amount of time they were spending on the phone.

7 A mechanism similar to the one described here, called a "scramble," is still used on some telephone circuits, especially those in cars.

8 It is difficult to trace the exact form this change might have taken. For one thing, it would have meant that subscribers could be connected, without Bell's intervention, with a wide range of people. Further, since private lines were so expensive, Bell might have been able to spend the money saved on wires on other features of its system or on a more equitable distribution.

9 In the early era of telephone development, telephone and telegraph poles were indiscriminately called "telegraph poles."

10 For more information on this issue, see BCA, ncm, 1881a, 1881b.

CHAPTER TWO

1 In fact, the very first exchange in the British Empire, according to Bell, had been opened in 1878 in Hamilton.

2 This meeting was to give managers from different areas a chance to get acquainted so that, in the future, they would be able "to meet the enemy with an united front" (BCA, d 26606, 1887, 43).

3 The People's Telephone Co. served an area which Bell Telephone Co. had refused to serve at first, except for Sherbrooke, the major city. However, when Bell saw the success of the independent company, it used all means at its disposal to "kill" it and take over.

4 For more information on prices and opposition, see BCA, d 1179, 1880; d 1180, 1880; d 1181, 1880; d 29867, 1880; d 24894-70, 5 July 1893.

5 Sise pointed this out in a letter to A. Robertson, president of the company (BCA, sle 1880).

6 In July of 1877, A.G. Bell assigned 75 per cent of his Canadian rights to his father, Melville Bell, of Brantford, Ontario. On 18 August 1879, M. Bell wrote T. Swinyard, managing director of Dominion Telegraph Co. in Toronto, offering to sell his rights to the company. Having received no reply, he wrote a second letter, and then tried to find other Canadian capitalists to whom to sell the firm for one hundred thousand dollars.

However, those who were interested in the company could not raise enough money to purchase it, and those who had the money did not wish to acquire such an uncertain enterprise. Thereupon, A.G. Bell urged the American company to purchase all of his father's Canadian rights (BCA, sb 80145b, 3146–2, 1958, 1–3). For more information on this issue, see Bruce 1973, 282–4.

7 Indeed, when Bell Telephone Co. was incorporated, in 1880, Sise's intention was to purchase all of the competitive telephone companies, especially those in big cities such as Montreal and Toronto. However, he raised only 70 per cent of the funds he required; the American company provided the remaining 30 per cent. In subsequent years, the percentage of Bell Telephone Co. of Canada stock held by the American company was as follows: in 1880, 24.8%; in 1885, 48.8%; in 1890, 47.8%; in 1895, 38.9%; in 1900, 38.5%; in 1910, 38.6%; in 1920, 38.3%. In addition, Sise himself was a wealthy American capitalist and, when the company had unexpected expenses mostly due to the "excessive cost" of line building or exchange fires, he lent Bell money (BCA sb 80145b, 3146–2, 1958, 3–4).

8 Letters from Sise to Vail, general manager of the American company, reveal this type of co-operation. See BCA, d 92711, 1885; d 93218, 1885. Similar letters are also available in Bell Canada archives. The first instrument designed and manufactured in Canada was made in the mid-1960s.

9 A station was "an installed telephone set with the associated wiring and apparatus" (BCA, d 29806, 1954). The number of stations did not correspond to the number of subscribers, as some users had more than one station in their household.

10 The increase was not that straightforward, however. Several people asked to have their telephones removed after they had tried them out for a time. As a result, the net increase over the period did not correspond to the number of people who were in contact with the new form of communication.

11 In a letter to H. Neilson, superintendent in Toronto, T. Swinyard, general manager in Toronto, explained that law firms were good customers to secure, as they were "willing to pay a good rental for such service as a telephone line would give them" (BCA, d 1127, 1879).

12 Residential charges in other areas varied from twenty-five to fifty dollars. When Bell Telephone Co. did not want to provide service to a particular class of subscribers, said the Toronto *Globe* in 1905, it charged exorbitant prices (*Globe*, 31 March 1905).

13 For example, in Montreal in 1887, a new exchange equipped with the multiple switchboard comprised two thousand lines (BCA, d 10900, 1887b).

14 From 1880 to 1890, "net earnings" grew from $724,378 to $5,260,712 (BCA, slo 7, 1892, 78).

15   Between 1899 and 1903, long-distance revenue increased by 114 per cent (BCA, slo 18, 1903, 63).

16   In 1905, in large Canadian cities, the ratio of telephones to inhabitants was about one to twenty, while in rural communities there was one telephone per 1,246 inhabitants (BCA, ncm 1905c).

17   For more information on the development of long-distance telephone networks, see Ogle 1979.

18   This was even more time since the regulation requiring Bell to connect independent telephone companies to its long-distance lines was not passed until 1912. For more information on regulations governing development of the long-distance telephone system, see Babe 1988.

19   The same calculation problem occurred in the telephone statistics published by the Dominion of Canada, starting in 1911.

20   The majority of subscribers were Anglophones. A telephone manager claimed that French-Canadians did not understand the use of the telephone. It was a cultural difference, he explained, as Francophones were more conservative than Anglophones. The disproportion in the number of agents working in Ontario and in Quebec might better explain the difference in numbers between English- and French-speaking subscribers. Moreover, the disproportion between the percentage of Anglophones and of Francophones in the ruling classes should also be taken into consideration. Finally, the fact that service was provided almost exclusively in English might have had some influence as well.

21   The provision of free telephones to some widely distributed newspapers was clearly based on discrimination by status, with the result that newspapers that were too small to warrant a free instrument were also generally too poor to buy one. It might have been expected that, in the long run, the lack of faster transmission by telephone would be detrimental to these small papers. It may be worthwhile to investigate the relationship between telephone distribution and the shutting down and starting up of newspapers during this period.

22   Ewen and Ewen accurately describe the goal behind such messages: "Mass imagery such as that provided by AT & T [AT & T and Bell Telephone Co. of Canada had similar advertising policies and used the same slogans] creates for us a memorable language, a system of belief, an ongoing channel to inculcate and effect common perceptions, explaining to us what it means to be a part of a 'modern world.' It is a world defined by the retail [individualized] consumption of goods and service ... a world where it increasingly makes sense that if there are solutions to be had, they can be bought" (1982, 42).

23   The first advertisement for telephone service featuring a photograph of a working-class person appeared in the early 1940s.

24   This was in great contrast to the telegraph, which was put under state management early in its development.

25   Competition was not strong during the early period of development, however, since Bell had acquired, by charter, all of the telephone rights and patents in central Canada – except Duquet's, which they bought very early – between 1880 and 1885. Moreover, subscribers were few, and wealthy, and did not contest telephone prices, although they often complained about the use value of the telephone.

26   This will be discussed later.

27   Another reason for Bell's approval of the commission, however, was that the company felt that the regulatory body had been created to limit not only Bell's activities, but also the activities of popular groups that were demanding nationalization of the company. According to Amstrong and Nelles (1986), Bell saw this political intervention as a "moindre mal" (lesser evil).

28   The secretary of the St. Jean-Baptiste workers' union stated that *La Presse* was "the true voice of the people," and added, "Let us praise it" (BCA, ncm 1918e). All quotations from *La Presse* have been translated by me.

29   As I stated earlier, the exception was the years 1914–1918, during which time the company did not encounter strong opposition to its rate increases.

30   For sampling of newspaper articles on the battle for rate increases, see BCA, nct 1913e, nct 1914a, nct 1915c, nct 1916a, nct 1918c, nct 1919a, nct 1920e.

31   This situation occurred although, as early as 1899, the company had included in its regulations a rule that allowed it to refuse service to "houses of doubtful or bad reputation" (BCA, d 811–2, 1899).

32   This elusive characteristic, so convenient to illegal businesses of that time, is also to be found in new technologies such as the cellular telephone, particularly used in cars, which seems to cause the police some problems in terms of criminal detection.

33   Mulock Commission, vol. 1, 622, quoted in Smythe 1981, 142.

34   Bell's major argument against government ownership was that non-users of the telephone should not be taxed to pay for telephone service which would benefit only a small group of people (BCA, d 12016, 1902; tg 1 (4) 1904). The business world carefully listened to and agreed with this stand. In effect, however, the cost of the telephone was assumed by the general population, as the prices of other commodities – food, clothes, and so on – were affected by increases in telephone rates. A state-owned telephone system would, according to some experts, mean lower rates, allowing a larger segment of tax-payers to afford it.

35   Sise's entries in his log-books noting the favours accorded to these politicians and the petition for posponement of government regulation of Bell

were close enough in time to suggest a link between favours, similar in nature, accorded on each side.

36   Bell, which had a monopoly on long-distance lines, refused to give other telephone companies the right to connect their systems to its lines.

37   The voided patent concerned only the telephone receiver, and not the transmitter, which was the most important part of the instrument.

38   Babe states that "between 1892 and 1905 in Ontario alone some 83 independent companies came into existence" (1988, 17).

39   For a list of doctors' telephone companies between 1890 and 1925, see BCA, d 29911, nd; d 29912, 1961; d 29913, nd; d 29914, nd.

40   See BCA, ncm 1918i, 1919h; nct 1913g, 1920a.

CHAPTER THREE

1   See, for example, Danylewycz and Prentice 1986; Grumet 1981; Prentice 1975.

2   See Lowe 1980.

3   See Briskin 1980; Curtis 1980; Luxton 1980.

4   In a capitalist patriarchal society, women's occupations are characterized by low wages and low-status jobs, and require such characteristics as obedience and submissiveness.

5   This periodical offered advice on how to economize, low-budget recipes, and so on.

6   In contrast to female operators, the prospect of advancement for male operators involved other positions, such as technicians and managers.

7   Shortly after Bell Telephone Co. started to hire women as operators, other independent telephone companies followed suit.

8   These "unsuitable" characteristics, however, might have been intensified by the low wages paid to operators. According to C.F. Sise, men would apply to be operators only "as a spare-time job as the income was too low" (BCA, sle 1887). Since they had other possible sources of income – even within the telephone industry – they were not particularly attracted to operator's work.

9   This worker was probably a night operator, as day operators were already women, especially in large cities.

10   Other occupations involved one or the other of these two characteristics. However, the case of operators was unique in that the job became a female occupation *uniting* them.

11   Dunstan and Baker were local managers for Bell in Toronto and Hamilton, respectively, and Mr Sise was general manager.

12   I use the word "girl," despite its sexist connotation, because it was how female operators were referred to by most people, including themselves.

13 At that time, women could also be hired to work on typewriters, replacing male copyists and shorthand writers. See BCA, ncm 1879.

14 Later, Bell Telephone Co. required five letters of recommendation. In addition to the three required previously, one letter was required from a physician and one from a property owner. *who were not suitable because of manners, & maturity etc.*

15 Some of these young men later became technicians, linemen, and even managers for Bell Telephone Co.

16 Since women were considered useless and inefficient in technical and mechanical work, some companies' managers had serious reservations about hiring them (BCA, d 27344, 1887, 5). These, however, faded when some female operators, especially in small towns, effected – with the aid of the instructions in the small book – all of the work necessary to maintain the telephone exchange in working condition.

17 Here I do not completely agree with Sangster (1978), who stated that the 1907 operators' strike in Toronto was the significant factor in bringing about more comfortable working conditions for the operators. Indeed, by 1896, the largest exchanges were provided with an "early form of air conditioning, recreational and lunch rooms, steel locker for each worker, drying room (even for clothing), bathrooms where last fashion baths were available for operators if necessary" (BCA, bp, 1980, 12). A telephone-company manager stressed that these improvements in the physical environment were not primarily for the operators' well-being: "I suppose the plan has its humanitarian side, but it is also a good business investment. Give your employees a pleasant and comfortable place in which to work, treat them with consideration and the results will more than repay for the trouble" (BCA, ty 8 (2) 1904, 124). This shows that, many years before the Toronto strike, telephone companies already understood some of the psychological principles behind employee productivity.

18 This story was told in 1937, which sheds more light on what she meant by "modern operator."

19 Most of the local managers were shareholders in Bell Telephone Co.

20 She was in charge of the operators for all shifts – day, evening, and night – although she worked only during the day shift. During evening and night shifts, the operators, who were still very few in number, had to perform some of the chief operator's duties: take note of long-distance calls, fill out complaint forms and so on. They gave her their reports in the morning.

21 A company manager in the "traffic department" clearly made this point in an article he published in the *Telephone Gazette:* "The object of the rules is to minimize operating labour and lost circuit time (thus loss of capital) and therefore, if they are disregarded, they are worse than useless for they only serve as a basis for argument" (BCA, tg 2 (9) 1911, 4).

22 In the United States, the first one was opened in New York.

23 For a discussion on the phenomenon of reification in relation to working-
    class consciousness, see Lukacs 1960, 110–41.

24 These were considered to be the characteristics of a good father.

25 If the subscriber was a prominent citizen, however, it was likely that the
    operator would hesitate before reporting him or her.

26 It seems that this type of infraction became so serious that C.F. Sise had to
    write to his local manager in Toronto to stress the "abuse of lines by
    employees" (BCA, slo 16, 1901, 9).

27 Although the petition did not say who the "fellow employees" were, it is
    safe to speculate that they were day operators in Toronto. I have not found
    any evidence of a group of operators making a comparison between their
    wages and those of either linemen or operators in another city.

28 In January of 1918, Mr W.G. McAdoo, Secretary of the Treasury in the
    United States, substantially increased the wages of American railroad
    workers to avoid any labour unrest during the war, and compensated for
    the increase by increasing freight and passenger fares by 25 per cent.
    Soon after, the Governer-in-Council in Canada ordered the "McAdoo
    Award" to be applied to publicly owned railway workers, and suggested
    that private industry follow its lead (BCA, d 18919, 1963, 46–7).

CHAPTER FOUR

1 "Over the telephone, the offensive voice cannot be softened or corrected
   by a glance or a smile," said a voice instructor from Emerson College of
   Oratory in Boston (BCA, ty 9 (5) 1905, 428).

2 Women were seen as "naturally" adaptable, and as promptly ready to
   follow instructions. These were considered to be biological characteristics.

3 There was no similar analysis of the Canadian voice but, since Canadians
   were second after Americans in frequency of phone use and also were
   settlers struggling against harsh conditions, it is safe to say that the telephone would have had the same effect, if any, on their voices as on those
   of their neighbours.

4 There's a tale from over the ocean of a deed that was nobly grand,
   But I would that the story of it might be writ by a worthier hand.
   How a cloudburst came to Folsom, but half the town was saved
   by the telephone operator who its awful fury braved.

   'Hello! Hello! Is that Folsom?' The call came clear that day:
   'Fly for your life to the mountain, a cloudburst comes you way,
   Passing us close but travelling along at a fearful rate
   To Folsom. You've twenty minutes, not more as I calculate,'

She sprang to the window, gasping 'Have mercy, gracious Lord.'
Below lay the quiet township. She turned again to the board.
With trembling voice and fingers she called them one by one
Till forty men in Folsom were bidden be up and run.

'Hello! Hello! D'you hear me? Up if you value your lives,
A cloudburst's coming over. Off with your children and wives.'
And the fortieth man had answered, she rang up forty-one –
But the storm had struck the cottage and her noble work was done.

The storm had passed. Day and night they are searching with bated breath
For the girl who died in harness at her post, a soldier death.
Twelve miles below in the valley they find her cold and dead,
Her battered and bent receiver still fastened upon her head.

She was only an operator, no heroine out of a book,
But Folsom will long remember the name of brave Sarah Rooke.
(BCA, qa 1909).

CHAPTER FIVE

1 Here I am considering only class differences, but there is also an interesting ethnic question at issue.

2 This percentage is deceptive, as there are only partial statistics on labourers in the "services" sector. A more accurate estimate would place the working classes at around 80 to 85 per cent of the population (deBonville 1975, 223). The description of the social distribution of the population in Montreal is based on DeBonville 1975 and Copp 1974.

3 From Ames 1897.

4 Canada Census, 1890–91, vol. IV, 283. Quoted in DeBonville 1975, 33.

5 Jean-Baptiste Gagnepetit was the pen name of a journalist at *La Presse* who had a daily column from 1884 to 1905 – with a brief hiatus in 1896 – and who was a strong denouncer of working-class exploitation, and of the terrible conditions in which working-class families were living.

6 From Yves Roy, *Alphonse Desjardins et les Caisses Populaires* (Montreal: Fides 1964). Quoted in DeBonville 1975, 110.

7 Report by Dr Louis Laberge on hygienic conditions in the City of Montreal. Quoted in DeBonville 1975, 127.

8 I obtained these figures by counting subscribers in the March, 1880, *Montreal Directory* (MMA 1880).

9 Although this is a 1924 map, the population distribution was about the same in the centre of the city. The difference was that in 1905 there were

fewer people in the northwest and centre-north sectors than there were in 1924.

10 See DMA, *Rapports pastoraux*, 1843–1935.

11 There was a discount of ten dollars per annum for doctors, though not for dentists or veterinary surgeons (BCA, d 29915–4, nd).

12 These included "Sister of Charity," 1898; "Tuberculosis," 1903; and "Charitable Refuge for Old Women," 1903. The charitable subsidies were few, and without exception were granted on the basis of a request either by the organization itself, or by a supporter (BCA, slo 13, 1898, 43; slo 18, 1903, 66).

13 Wooden sidewalks, as well as the wooden sewage system, were said to be very unsanitary. The whole system was infested with rats, which did not discriminate according to class. Therefore, under pressure from the ruling classes, the city built seventy-five miles of concrete sewers between 1872 and 1882 (MMA 1942).

14 See DMA, *Mandements, lettres pastorales, circulaires et autres documents.* Here are some examples of the archbishop's recommendations: "I forbid in the strongest terms possible these excursions [picnics] and pleasure parties" (*Circulaire de Mgr l'Évêque de Montréal au clergé de son diocèse* 29, 30 May 1880). He also asked priests in all parishes to "bind your parishioners to refrain from frequenting suspect and improper theatres" (*Circulaire de Mgr 68*, 7 June 1885). Finally, he strictly forbade parents to send their children alone to balls (*Mandements* 13, 16 December 1901). Although members of the English bourgeoisie generally did not practice Roman Catholicism, their lives were also based on strong religious principles. See Haller and Haller 1974.

15 Unfortunately, there are no sociological or historical studies that discuss visiting patterns during the period 1870–1920. Biographies and novels provide the best data, although some information can be found in Fahmy-Eid and Dumont 1983; Katz 1975; and Miner 1939.

16 This information was gathered in an interview with John Radcliffe, of Bell Canada.

17 Telephone numbers were not yet in use.

18 This instruction was necessary only after 1884, when calls were made with numbers instead of names.

19 Those who were young enough to play with the telephone were too young to read the instructions.

20 See BCA d 12016, 1879a, 1881.

21 Unless unexpected elements interfered with the transmission. For instance, *Nature* reported in 1879 that the Edinburgh Liberals, who were fond of science and technology, decided to transmit a political speech by one of their heros, "Mr Gladstone," over the telephone so that he "might reach a

much larger audience." Unbeknownst to Gladstone, there was a telephone in the hall where he was speaking. However, "he put his hat on the cylinder of the telephone and the communication was cut for most of the speech" ("Notes," 1879, 115).

22 The newspaper did not use a regular telephone line, but extra lines that were not being used by the telephone company. The company tried to cancel the deal before transmission started, but without success. However, it cancelled the lease after a year, which was why the newspaper did not last longer. This was a one-way communication, and the only way that the subscribers could show their dissatisfaction was by hanging up or by banging the receiver on the wall! For more information see Marvin 1988, 233–31.

CHAPTER SIX

1 Fischer (1988a) situates the rapid growth of rural telephony by independent companies in United States at around 1893, after Bell's patents expired. For more information on American rural telephone-system development, see Fischer 1988a.

2 No secrecy switch of any kind was promoted by Bell Telephone Co., or by any other company for that matter, whereas private lines were strongly recommended, in the very first advertisement, as something which could be built "on reasonable terms" (BCA, d 12016, 1885b).

3 Automatic telephones had been marketed earlier in Paris and New York.

4 He did not say, however, what pressure was put on the operator if she refused to do it.

5 For more information on this issue, see BCA, ncm, 1918a.

6 There was another lengthy debate on the right of police forces to tap telephone wires in the late 1960s, when John Turner was justice minister, and a more specific bill was enacted by Ottawa to limit police activities.

7 For additional reports on invasion of the home by telephone, see BCA, ty 1904, 7 (4), 309; ty 1905, 10 (6), 433.

8 Fischer situates the beginning of the American telephone companies' advertisement of the telephone for sociability at the end of the 1920s. For more information on American development, see Fischer 1988b.

9 The agenda was drawn up by an operator.

10 See BCA, ty 1905, 10 (3), 211; ncm, 1908a; 1919a. Marvin also makes the point that men in the telephone business thought that "women failed to understand electrical messages the way their male protectors did, as scarce and expensive commodities," and that "their [women's] conversation [was] trivial and uninformative, and could [have been] easily managed face-to-face." Marvin 1988, 22–32.

11 The home, however, was itself becoming less private with the advent of the telephone. For more information on the danger of the bicycle for Victorian women, see Haller and Haller 1974, 174–87.

12 On the other hand, Fischer argues that in United States rural residents were independent and had very little intercourse with their neighbours. See Fischer 1988b.

13 It is interesting to note that, a few years after Bell Telephone Co. had, with great effort, eliminated party lines in cities and built its system on the basis of private lines only, with the hope that one day it would be pre-eminent even in rural areas, it reintroduced, for an extra fee, a service with the advantages of the party line. The "telephone conference service," started in the early 1930s, was available for business and for "social use" (Banning 1936, 146). One difference between these services lay in the distance they covered. Party-line service was limited to local calls, whereas conference service was available for long-distance communication. Still, it was possible to adapt party-line service to long-distance service. Another important difference was that "meetings" on party lines did not involve as exclusive a group of callers as did "conference calls," since any subscriber connected to the party line could listen to or participate in conversations, whereas conference-call participants were predetermined.

14 Statistics on telephone conversations for Canada are not available. However, as figure 10 shows, Canada was only slightly behind the United States in terms of telephones per capita – 10 per cent of the population in Canada in comparison to 12 per cent for the United States – and Canadians had a reputation for being heavy telephone users. Consequently, the figures for telephone conversations for the United States give a good idea of what was happening here. Development of the telephone in the United States was generally comparable to that in Canada, as one might expect, since many factors were similar: the same company, with management in continual contact, same types of population, and so on. There were, however, some variations, as Fischer (1988b) points out.

15 See also Rhodes, 1927.

16 She was probably afraid of being caught by her employer. Domestics usually were not allowed to use their employers' telephone for personal calls.

17 The law started to be applied early in 1916. See: BCA nct 1916d; 1916e.

18 As late as 1905, Bell managers were still writing to each other, instead of telephoning, for business matters. See: BCA, sle 1905a, b. Although they complained about the poor postal service, they continued to do business via correspondence, even to locations within telephone reach. Was the telephone too indiscreet for them – or perhaps, too expensive?

19 See BCA, d 12016, 1880e; qa 1880e, 1880d, 1914.

20 Shaw worked for a British telephone company for some years at the beginning of his writing career.

21   Here are some samples: "Telephone and Germs," *Montreal Star,* 11 Sept. 1905; "New Way to Telephone," *Montreal Gazette,* 24 Jan. 1907; "The Telephone and Microbes," *Montreal Star,* 30 July 1908; "Germs in the Telephone," *Telegram,* 13 Jan. 1916; "Germ Proof Phone," *Herald,* 17 Feb. 1916.

22   As we saw earlier, the operator was instructed to ring a subscriber no more than twice.

23   Blushing, which was Victorian women's general expression of embarrassment, was, of course, not discernable over the telephone. Embarrassment was thus expressed in vocal nuances, which were imperfectly transmitted over the telephone.

CONCLUSION

1   However, the replacement of operators with automated switchboards was related to other conditions of development. One of them was competition. In central Canada, for instance, Bell had little technological competition until the end of 1919, when its patent expired. It was only then that the company considered automatic telephony. In other cities, in western Canada and the United States, automatic switchboards began to be used as early as 1905.

2   These practices are reminiscent of Burawoy's (1979) notion of "making out" developed in *Manufacturing Consent.* However, although they were the same kind of insider rules, the operators' opposing practices did not stimulate worker productivity. Rather, because their role was that of mediator between the company and the subscribers, their resistance embarrassed the former and enraged the latter.

3   However, early in the development of long-distance communication, some women were conscious of its liberating possibilities. Indeed, American feminists in the 1850s asserted that "women, rather than men, must control these new services [the telegraph] and use them as their base of economic power" (Hayden 1981, 12). They considered telegraph to be not only a useful means of communication, but also a business that could provide them with the funds necessary to improve their domestic conditions.

4   They can also be race oriented, but this is an issue that I chose not to discuss here.

Bibliography

ABBREVIATIONS

BCA Bell Canada Archives
  nd No date
  bb *Blue Bell*
  bg Background, 1870, 1880, 1885
  bp Bell publication
  d   Document
  hit Highlights and illustrations, Toronto, 1877–1925 (four files)
  ls   Life story
  me Montreal events, #1–3
  nca Newspaper clippings, advertisement
  ncm Newspaper clippings, Montreal
  nct Newspaper clippings, Toronto
  qa Quotations and anecdotes (six files)
  sb Storage box
  sf Subject file
    a   advertisement.
    il illegal use of the telephone.
    in invention of the telephone. (two files)
    ind invention and development of the telephone. (two files)
    me Montreal exchanges (seven files)
    op operators. (two files)
    sec secrecy
    ts telephone service
    ty traffic years, 1901–1916
    wo women
  sle C.F. Sise Letters, 1880–1882
  slo C.F. Sise log books, #1–20, 1885–1905
  td Telephone directory
  tdm Telephone directory Montreal
  tdt Telephone directory Toronto
  tg *Telephone Gazette*
  ty *Telephony*
DMA Diocese of Montreal Archives
MMA Montreal Municipal Archives
MML Metro Montreal Library

ARCHIVAL SOURCES

Bell Canada Archives

BLUE BELL

1924   "Le téléphone automatique," Jan., 15.
1930   "Half a century progress," Jan.
1943   "Le calculagraph," July, 41.
1950   "Le téléphone," 93.
1951   "In talking long-distance," Jan.
1959   "Our panorama drugstore," Apr., 100.

BACKGROUND

1880   "The telephone," *Toronto Globe*, 18 Sept.
1885   Letter from C.F. Sise to T.N. Vail, 8 May.

BELL PUBLICATION

1980   *One Hundred Years*. H.G. Owen. Montreal: Bell Canada.
1981   *Alexander Graham Bell*. Montreal: Bell Canada.

DOCUMENTS

803   1890   Montreal Telephone Directory, August.
811–12   1899   "The Bell Telephone Co. of Canada (Ltd), Rules and Regulations," Jan.
896   1880   "Official List of Connections of Bell Telephone Exchange," March.
920   1892   "Rules and Instructions for Operators." Montreal: The Bell Telephone Co.
926   1899   "Toll Line Rules." Montreal: The Bell Telephone Co. of Canada Ltd., 1 July.
1009   1934   No title. *Telephony*, 1 Sept.
1059   1879   Letter from a subscriber to Wadland, 20 Dec.
1065   1879   Lease to C.D. Cory and H.C. Baker, Hamilton, 18 Oct.
1069   1880   45–46 Victoria (1880) Ch. 71, Sec. 2. "The Bell Telephone Co.: Act of Incorporation."
      1881–1882   "The Bell Telephone Co.: Act of Incorporation, Provincial Enabling Acts" (Quebec, 1881; Ontario, 1882).
1069–4   1902   2 Edward VII (1902) ch. 41, sec. 2. "The Bell Telephone Company: Act of Incorporation and Amendments."

1127	1897	Letter from H. Neilson to T. Swinyard, 22 Jan.
1173	1880	Letter from L.B. McFarlane to T. Swinyard, 9 Feb.
1179	1880	Letter from J. Stewart to L.B. McFarlane, 24 Feb.
1180	1880	Letter from P.W. Snider to L.B. McFarlane, 23 Feb.
1181	1880	Letter from G.P. Dunlop to L.B. McFarlane, 24 Feb.
1239	nd	"Office Management. Charges for extra mileage. Joint use of telephone. Movings. Ect. Maintenance of friendly relations with general public." K.J. Dunstan, general manager, Toronto.
1277	1982	Letter from C.F. Sise to A.W. Barnard, 24 Mar.
1544	1898	Advertisement, no source.
1544–2	1898	Bell's long-distance telephone system in Central Canada.
1857	1916	"Courtesy." C.F. Sise, general manager B.T.C., 1 May.
1978	1920	"Announcements," Bell Telephone Company.
2405	1880	K.J. Dunstan's payroll, Oct.
2935	1903	"Statistics." Bell Telephone Company.
3168	1890	"The first underground cables in Montreal." No source, Aug.
3379	1926	"Lecture on the telephone business: The regulation of telephone rates." J.E. Macpherson.
3521	1899	"Montreal Exchanges."
3653	1882	"The Electric Despatch Co. Toronto." 16 Jan.
6077	1903	Letter from A. Meighen Bros. to Gilman. 1 Sept.
6453	nd	"The inviolability of telegraph." No source.
6454	nd	"The telegraph and the telephone in this city." No source.
6458	nd	Newspaper article. No title, no source.
6680	1905	"A successful independent telephone company." C. Skinner. *Canadian Engineer*, Feb.
7925	1888	Letter from Baker to C.F. Sise, 20 Jan.
7934	1887	Letter from Coulson to Baker, 3 Sept.
7966	1892	Letter from Baker to C.F. Sise, 1 Jan.
8162	1879	Lease to J.R. Lee, Toronto, 18 Dec.
8353	1918	"Complaints." *Plant Bulletin*. Montreal: Bell Telephone Co., #48, 28 June.
9126, 9128, 9129	1885	Series of letters from C.F. Sise to Baker, 19, 22 Sept.; 22 Nov.
9498	1881	Letter from H.L. Powis(?) to Bell Telephone Co., 16 Dec.
9710	1888	Letter from C.F. Sise to Smith, 7 Nov.

10900	1887a	"Northern Electric Co. history." C. Schoebly.
	1887b	"Multiple Magneto Switchboard."
12015	1877a	"Local test." *Expositor*. 1 Sept.
	1877b	"Telephone entertainment." *Evening Times*, Nov.
	1877c	"The telephone unmasked." *The Peterborough Times*, 2 Nov.
	1881	"The Bell telephone." *Daily Star*, 26 July.
	1882	"A secret telephone." *Daily Star*, 25 Apr.
12016	1879a	"A sermon by telephone." *Evening Post*, 2 Dec.
	1879b	"Success of the telephone." *Peterborough Review*, 19 Dec.
	1879c	"The Dominion Telegraph Co., district telephone department."
	1879–1906	L.B. McFarlane's scrapbook.
	1880a	"Prescription by telephone." *Daily Herald*, 10 Jan.
	1880b	"The right of citizens must be respected." *Quebec Telegraph*, 15 Dec.
	1880c	"Telephone exchanges." W.H. Harper.
	1880d	"Letter from people." W.H. Harper. No source.
	1880e	"Still Single." No source.
	1880f	Advertisement. No source.
	1881	"An eighth wonder." Toronto *Mail*, 24 Oct.
	1884a	"Quiet telephone." No source.
	1884b	"Hilloa! Hilloa!" No source.
	1885a	"Telephone competition." *Montreal Star*, 27 Jan.
	1885b	Advertisement. No source.
	1892	No title. *Daily Star*, 7 Oct.
	1898	"Who wouldn't be a telephone girl?" *Watchman*, 12 June.
	1902	"The tax-payers." No source.
	1906	"The wonderful development of the telephone in Canada." *Montreal Witness*, 20 Jan.
	nd	"Across the wires." *Montreal Witness*, 21 Mar.
12113	1878	No title. *Daily Expositor*, 4 Feb.
12126	1878	No title. *Daily Expositor*, 18 Nov.
12267	1884	Letter from C.F. Sise to R. Reid, 30 Jan.
12444	1952	"All our yesterdays." Montreal *Gazette*, 23 Aug.
12451	1879	"Hamilton letters," 1 July.
14220	1907	*Report of the Royal Commission on a Dispute Respecting Hours of Employment between The Bell Telephone Company of Canada, Ltd. and Operators at Toronto, Ont.* Ottawa: Government Printing Bureau.

17958	1905	*4–5 Edward VII, A, 1905. Report of the Select Committee on Telephone Systems*, minutes of evidence, appendix 1, 805–6.
18919	1963	*A History of Labour Relations in The Bell Telephone Company of Canada, 1880–1962*, G.M. Parsons. Montreal: Bell Canada Publications.
19406	nd	"Changing years as seen from a switchboard." R.T. Barrett. No source.
21203–2	1908	Advertisement. *The Farmer's Advocate*, 17 Dec.
21732–8	1881	Letter from Duquet to C.F. Sise, 7 Jan.
21732–9	1881	Letter from C.F. Sise to Duquet, 11 Jan.
24089	1888	"Use of women operators at night." *Quebec City Letters.*
24092	1914	"Circular." From L.B. McFarlane to Bell superintendents and managers, 17 Aug.
24096	1884	Letter from L.B. McFarlane to L. Irwing, 12 Jan.
24894–2	1882	Letter from L.B. McFarlane to R.N. England, 16 Jan.
24894–70	1893	Letter from C.F. Sise to J.E. Hudson, 5 July.
26606	1887	Report of meeting of C.F. Sise with local managers. 16–17 May.
27144-44	1887	Letter from L.B. McFarlane to C.E. Cutz, 7 July.
27344	1887	Questionnaire from C.F. Sise to local managers, 16 May.
28012	1880	Photograph. No source.
29144–38	1888	Letter from C.F. Sise to E.P. Lachappelle, 31 Mar.
29144–53	1884	Letter from Sclater to F.G. Walsh, 16 Jan.
29144–54	1884	Letter from Sclater to M. Seaborn, 16 Jan.
29516	nd	"Rural party line telephone service: abuses and annoyances and their remedies." The Eastern Township Telephone Co.
29547	nd	Newspaper article. No source. No title.
29547–1	1904	Letter from Taylor Bros. to Bell Telephone Co., 3 Feb.
29806	1954	*Glossary of Telephone Words and Terms.* E.C. Smith.
29867	1880	Letter from H.C. Baker to L.B. McFarlane, 18 Oct.
29870	1880	Letter from C.F. Sise to H.M. Dougal, 3 Sept.
29909	1967	"The life-saving line of a country doctor." E. Lunney. *Family Herald*, 2 Feb.
29911	nd	"Doctors Organising Telephone Companies."
29912	1961	"Patient's death inspired rural telephone network." *London Free Press*, 18 Feb.
29913	nd	"Doctors Organising Telephone Companies."

29914   nd   "Doctors who Played Prominent Role in Early Telephone Service."
29915–3   1902   Letter from L.B. McFarlane to Sclater, 30 Apr.
29915–4   nd   "Rates."
29920   1879   Advertisement. *Montreal Star,* 1 Nov.
30114   1965   "The druggists 'dread.'" *Montreal Gazette,* 11 Sept.
30744   1880   Letter from Baker to C.F. Sise, 4 Aug.
30753   1885   No title. Toronto *Globe,* 27 Apr.
92711   1885   Letter from C.F. Sise to T.N. Vail, 8 May.
93218   1885   Letter from C.F. Sise to T.N. Vail, 18 May.

HIGHLIGHTS AND ILLUSTRATIONS, TORONTO

1877–1909   "Toronto rates."
1877   "Police patrol system." *Telephone Gazette,* 2 (2): 9.
1895   "Once upon a time short skirts were a must." *Bell News,* 30 Oct.

LIFE STORY

1880   Dunstan, K.J., manager.
1880–91   Warren, M.R., operator, chief operator.
1880–95   Camp, L.W., operator, chief operator.
1883   Clarke, W.J., maintenance.
1884–91   Unidentified operator.
1888–96   Helsby, E.M., operator.
1889   Greendale, F., operator.
1889–93   Phillips, M.I., operator.
1892   Flanagan, M.F., operator.
1893–1912   Earle, W.D.V., maintenance, local manager.
1896   Chevalier, W., operator.
1899–1910   King, M.E., operator, supervisor.
1899–1911   Lasalle, H.J., operator, supervisor.
1900   Hill, M., operator.
1900–01   Kerr, D., operator.
1902   Stead, I., operator.
1906   Lalonde, B., operator.
1907   Cline, C.M., operator.
1907–17   Johnson, H.V., operator.
1908–17   Hannaford, E.V., operator, supervisor.
1910-13   McCutcheon, E.M., operator.
1913   Lachapelle, M.A., operator.

1917   Maloney, M.K., operator.
1918–20   Geach, O., operator.
nd   Bushell, A., operator.

MONTREAL EVENTS

1880   "First woman operator."
1885   *Montreal Gazette*, Oct.; Montreal *Minerve*, Oct.;
           Montreal *La Patrie*, Sept.
1894   "Lady operators benefit association." Constitution
           and By-Laws, No. 25763, Report, No. 28878.
1897   Letter from L.B. McFarlane to local manager,
           Montreal, 13 Aug. 1896.

NEWSPAPER CLIPPINGS, ADVERTISEMENTS

1903–13   Advertisements, *Sunday World*.
1911b   Advertisements, *Quebec Daily Telegram*, 9 Sept.
1911a   Bell advertisement. No source. 19 Aug.
1912   Bell advertisement. *Pioneer*, July, Aug., Sept.
1923   "Les gens qui demandent l'heure retardent le ser-
           vice." *Événement*, 15 Feb.
1925   "The 'voice' returned." *Telephone Almanacs*, Oct.

NEWSPAPER CLIPPINGS, MONTREAL

1879   "New Industry." *Montreal Daily Witness*, 4 Jan.
1880a   "Medico-chirurgical society." *The Free Press*, 10
           Jan.
1880b   *Star*, 13 June.
1880–1905 "Modern marvels of telephoning." *Montreal Star*,
           12 Jan.
1881a   *Quebec Daily Telegraph*, 17 Feb.
1881b   *Quebec Mercury*, 28 June.
1897   "Telephone rates." *Montreal Star*, 25 Mar.
1902a   "Telephone bill in sub-committee." *Montreal Star*,
           4 Apr.
1902b   "Biggest Canadian electric contract." *Montreal
           Star*, 12 June.
1905a   "Telephone inquiry." *Montreal Gazette*, 10 Mar.
1905b   "Telephone inquiry." *Montreal Gazette*, 23 Mar.
1905c   Advertisements. *Witness, Cigar and Tobacco Jour-
           nal, Le Canada, La Presse, Star, Gazette, La
           Patrie*, 24 Mar.
1905d   "Telephone inquiry." *Montreal Gazette*, 26 May.
1905e   "Telephone talks," *Montreal Gazette*.

1906a   "Toronto's telephones." *Montreal Star*, 28 Feb.

1906b   "Ontario's telephones." *Montreal Gazette*, 1 Mar.

1906c   "Two Canadians have solved problem of automatic 'phone.'" *Montreal Star*, 3 Mar.

1906d   "Brutalizing telephone." *The Graphic*, 3 Mar.

1906e   "The telephone supersede telegraph in rail roading." *Herald*, 25 June.

1906f   "Telephones as germ collectors." *Montreal Star*, 25 June.

1906g   "New resources of the farmer's wife." *Montreal Star*, 23 July.

1906h   "The rural telephone revolution." *Montreal Star*, 31 Aug.

1907a   "Tapping the wires." *Montreal Journal of Commerce*, 8 Feb.

1907b   "Telephone inquiry." *Witness*, 9 Feb.

1907c   "Telephone inquiry wages and living expenses." *Montreal Gazette*, 11 Feb.

1907d   "A natural situation." *Montreal Gazette*, 19 Sept.

1907e   "The diary of a telephone girl." *The Saturday Evening Post*, 19 Oct.

1908a   "Deplores gossip over telephone." *Montreal Herald*, 2 Apr.

1908b   "Train and telegraph." *Herald*, 25 June.

1908c   "Héros inconnus et méconnus: la téléphoniste." *La Patrie*, 14 July.

1908d   "Deadly deseases locked in telephone receivers". *Herald*, 8 July.

1908e   "Le téléphone." Tante Ninette. *Le Bulletin*, 6 Sept.

1908f   Advertisements. *Herald*, 26, 28, 30 Oct.; 2, 6, 9, 11, 15, 16, 18, 20 Nov.

1909a   "New phone exchange for north end will be inaugurated August 15." *Montreal Gazette*, 8 Aug.

1909b   "To switch 2,500 'phones almost instantaneously." *Montreal Star*, 14 Aug.

1909c   Advertisement. *Montreal Star*, 18 Sept.

1915a   "The etiquette of the telephone." *Evening News*, 11 Mar.

1915b   "Phones in restaurant." *Herald*, Dec.

1916a   "Voice culture." *Montreal Mail*, 2 July.

1916b   "Girls were never greater blessing than when they entered phone exchanges." *News*, 4 Nov.

1917a   "The singing central." *Herald*, 9 Mar.

1917b   "Fonds de pension pour les employés du télé-
         phone." *Le Canada*, 18 Apr.

1917c   "Un geste vraiment généreux." *La Presse*, 18 Apr.

1917d   "Telephone secrecy." *Montreal Gazette*, 5 Dec.

1918a   "'Listening in' on telephone to stop." *Montreal
         Star*, 28 Jan.

1918b   "Eavesdropping on 'phones and offence." *Montreal
         Gazette*, 31 Jan.

1918c   "Telephone courtesy." *Herald*, 16 Mar.

1918d   "Un ultimatum des pharmaciens à la Cie de télé-
         phone Bell," 23 Nov., no source.

1918e   "Le téléphone pour tous." *La Presse*, 26 Nov.

1918f   "Le peuple contre le trust qui l'exploite." *La
         Presse*, 27 Nov.

1918g   "Le téléphone ne doit pas être tenu comme un
         luxe." *La Presse*, 28 Nov.

1918h   "Des nécessiteux qui thésaurisent." *La Presse*, 6
         Dec.

1918i   "Le téléphone progressif." *La Presse*, 14 Dec.

1919a   "L'abus du téléphone." *La Presse*, 10 Jan.

1919b   "More 'houses' minus a phone." *Montreal Star*, 17
         June.

1919c   "Prohibition hurts 'phones". *Montreal Star*, 24
         July.

1919d   "Two phone operators." Letter to the Editor,
         signed "o dam." *Montreal Star*, 17 Sept.

1919e   "Why wrong numbers are given?" *Montreal Star*,
         25 Sept.

1919f   "Not always the operator's fault." Letter to the
         Editor, signed "Hello girl." *Montreal Star*, 29
         Sept.

1919g   "Usually the public's fault." *Star*, 4 Oct.

1919h   "Le mauvais service des téléphones." *La Presse*, 15
         Dec.

NEWSPAPER CLIPPINGS, TORONTO

1881   "Sunday and the telephone." *Globe*, 19 Nov.

1905   *Globe*, 31 Mar.

1911a   Advertisement. *News*, Apr.

1911b   Advertisement. *Chronicle*, Aug.

1912–13   Advertisement. *Globe, Mail*, Dec.

1913a   "Square deal for north, says Mayor." *World*, 21
          July.

1913b   "Poor central has very busy and hard life." *News*, 26 July.

1913c   "For telephone girls are educating subscribers." *Star*, 13 Sept.

1913d   "'Phones in Toronto double in 5 years." *Star*, 16 Sept.

1913e   "Sault and Steelton apparently have no say on 'phone rates." *Star*, 25 Oct.

1913f   "Telephone users tell their troubles." *Star*, 1 Nov.

1913g   "'Phone service good! Is the manager joking?" *Globe*, 3 Nov.

1913h   "'Phone Co's pact is ratified." *Mail*, 20 Nov.

1913i   "Speak nicely to the Hello Girl." No source.

1914a   "Evidence is wanted in telephone rates case." *News*, 18 Feb.

1914b   "Why Central never gives wrong numbers." *Star*, 7 Mar.

1914c   "Box alarms are quickest." *Telegram*, 28 Mar.

1914d   "Stupid 'Central' caused long delay." *Toronto Star*.

1915a   "Telephone manners." *Star*, 9 Mar.

1915b   "Toronto has no need to swear at Central." *Star*, 10 Mar.

1915c   "Reduce telephone rates." *Telegram*, 25 Mar.

1916a   "Reduce telephone rates." *World*, 11 Jan.

1916b   "Reduce telephone rates." *World*, 11 Jan.

1916c   "Used bad language over the telephone." *News*, 1 Feb.

1916d   "First conviction for foul language." *Star*, 1 Feb.

1916e   "The privacy of the telephone." *News*, 25 May.

1916f   "Voice culture." *Mail*, 2 Aug.

1916g   "Would reduce H. C. of L." *Telegraph*, 23 Nov.

1918a   "Doings of the Duffs." *Star*, 21 Mar.

1918b   "As a result of strike." *Star*, 23 July.

1918c   "To raise 'phone rates." *Star*, 4 Oct.

1919a   "Watch your 'phone bills." *Telegram*, 7 July.

1919b   "New device gives secrecy to phone." *Mail*, 7 Aug.

1920a   "Quite complicated." *Star Weekly*, 3 Jan.

1920b   "Woman badly beaten, man made his escape." *Star*, 24 Feb.

1920c   "Says 'machine girl' will give your number." *Star*, 18 Mar.

1920d   "K.J. Dunstan denies Bell service is bad." *Star*, 20 Mar.

1920e   "Discuss telephone rates." *Star,* 25 Aug.

1920f   "How increased phone rates would affect all citizens." *World,* 14 Dec.

1952   "Says telephone operator criticized unjustly." *The Windsor Daily Star,* 27 May.

QUOTATIONS AND ANECDOTES

1877   "The telephone at work." *Brantford Expositor,* 29 Sept.

1879   "A telephone feat." *Herald,* 24 Nov.

1880a   "The telephone." *The Morning Herald,* 9 Feb.

1880b   No title. *Blue Bell.*

1880c   "The telephone." *Punch,* London, England, no date.

1880d   "Yelling through the telephone." *The Electrician,* Aug.

1881a   "A rough ruffian." *Montreal Witness,* 23 Feb.

1881b   Letters from C.P. Sclater to F.G. Walsh, 12 Dec.

1884a   No title. *The Farmers Advocate,* 15 May.

1884b   "L.B. McFarlane's statements."

1903   "The telephone and the modern world." *Telephony,* 343.

1906   "Shades of 1906." *Flasher and Plugs,* 20 May.

1909   "A telephone heroine," E.M. Buckland. *National Tele-Journal,* London, England.

1914   "The value of the telephone." E.A. Guest, no source, 17 Nov.

1918   "The friendly phone," B.D. Fowler. *The Country Gentleman,* 30 Mar., 119–20.

1919   "The suffragette." *Montreal Herald,* 18 Feb.

1936   "The importance of a party line." *Globe and Mail,* 11 Aug.

1973   No title, no source.

1979   "Phone in vaults," J. Collins. *The West End Branch,* 21 July.

nda   "Telephone technical description."

ndb   "Special notice to party line subscribers." The Bell Telephone Company of Canada.

ndc   "Collinwood, Ontario."

STORAGE BOX NUMBER

9497.110,4167   1903   Letter from C.F. Sise to L.B. McFarlane, 6 June.

59497.110,4167   1903   Letter from C.F. Sise to L.B. McFarlane, 12 June.

80013c,3144–2   1880   "Telephone history – Montreal."

1887   "Montreal – Questionnaire," 23 Feb.

80041b,3146–3   1907   "Telephone service to Toronto people is growing worse." *News*, 20 Jan.

1938   "History of Toronto telephone development." G.C. Jones.

80141b,3146–3   1900   Letter from K.J. Dunstan to C.F. Sise, 19 Apr.

1902   Letter from K.J. Dunstan to local manager in Hamilton, 2 Jan.

80145b,3146–2   1883   "Memoranda re telephone development in Toronto since 1883." Head Office Incoming Correspondence.

1894   "Toronto exchanges."

1958   "Summarized telephone history of Toronto, Ontario." Bell Telephone Co. publication.

nd   "Special rates in Toronto."

80910b,3144–1   1881a   Letter from L.B. McFarlane to C.F. Sise, 23 Apr.

1881b   Letter from L.B. McFarlane to H. Trudeau, 30 Dec.

1892   "Montreal exchanges."

1914   "Total number of stations and operators."

84141b,3146–3   1902   Letter from K.J. Dunstan to Bell Telephone Co. management, 22 Jan.

86245,1587   1915–22   "Bell Telephone Co. of Canada in cities."

SUBJECT FILES

Advertisement

1878   "The speaking telephone." Circular issued by District Telegraph Co., November.

1900   Bell Telephone Co. advertisement. No source.

1902   Bell Telephone Co. advertisement. *District of Eastern Ontario Directory*.

1904   Advertisement. *Quebec Telephone Directory*, August.

Invention

1877a   *Revue et Gazette Musicale de Paris*, 14, 8 Apr.

1877b   "The telephone," G.G. Hubbard, May.

1881   Advertisements.

1896   "Some dates in Canadian telephone history."

nd   "A few facts about the creation of the telephone." *Electrical Talks* 30.

Invention and development

1877   Letter from C.D. Cory, 27 Sept.

1895   "Montreal telephone history." L.S. Moore.

1897   "Barb wire lines." *The Family Herald and Weekly Star*, 18 May.

1900   "The big change." F.L. Allen.

1904   Letter from L.B. McFarlane to W.J. Gilmour, 25 Jan.

1921   "The theatre, the drama, the Girls." G.J. Natham.

1929   "The telephone." *Bloor Watchman*, 30 May.

1931   "The subscriber who was lost," L.B. McFarlane. *The Canadian Magazine*, Jan.

1938   "This brave new world." Aug.

1945   "History of the telephone."

1952   "The development of the telephone." *Telephone Historical Collection*, Montreal, 8 Oct.

1977   No title. *Pioneer*, Aug.

1980   "The telephone and social change."

nd   "A few facts about the creation of the telephone." *Electrical Talks* 30.

       Illegal use of the telephone

1880a 44–45 Victoria Act (1880) Ch. 67 Sec. 25.

1880b 44–45 Victoria Act (1880) Ch. 67 Sec. 521.

1902   Brief of the Bell Telephone Company of Canada. Montreal exchanges

1890   "First long-distance line between Quebec/Montreal." 1 Sept.

       Operators

1897   "Instructions to operators at agency calling; and instructions to operators at agency called."

1930   "I was your old 'Hello Girl,'" K.M. Schmith. *The Saturday Evening Post*, 12 July, 18–19, 120–22.

1940   "Sixty years at the switchboard." *Michigan Bell Magazine*, Oct., 4–5, 21.

1941   "Heroines of switchboard have saved many lives by quick wit and bravery." No source.

nda   'Goodbye Central," L.G. Erdman. *Rural New-Yorker*.

ndb   "'Bread toaster' to common battery." No source.
       Secrecy

1881   Letter from L.B. McFarlane to A.B. Caldwell, 25 Oct.

1941   Extract from Telephone Companies Act (Quebec), R.S.Q. 1941, Chapter 298.

1950   Extract from the Telephone Act (Ontario), R.S.Q. 1950, Chapter 387.
       Telephone service

1927   "Forcasting the demands for telephone service," R.V. Macauly, 10 Feb.

Traffic years, 1901–1916

1910 "The history of the telephone," H.N. Casson. *Electrical Review and Western Electrician*, 9 July, 57 (2): 86–90.

Women

1951 "Permits for prerogatives," F. Tumpane. *Globe and Mail*, 28 Feb.

nd "Really 'Hello girl.'" No source.

C.F. SISE'S CORRESPONDENCE

1880a Letter from C.F. Sise to A. Robertson, 17 Jan.
1880b Letter from C.F. Sise to W.H. Forbes, 15 Mar.
1887 Letter from C.F. Sise to a local manager, 21 Feb.
1895 Letter from C.F. Sise to K.J. Dunstan.
1905a Letters from C.F. Sise to F.P. Fish, 1 Apr., 3 May, 16 June, 20 June.
1905b Letter from C.F. Sise to W.S. Allan, 30 June.

C.F. SISE'S LOG BOOKS:
1885–1905 Log Books Nos 1 to 20.

TELEPHONE DIRECTORY

1886 Halifax telephone directory.
1890 Hamilton telephone directory.
1906 Eastern Ontario telephone directory.
Montreal
1880–1920 Montreal telephone directories.
Toronto
1879–1920 Toronto telephone directories.

TELEPHONE GAZETTE

1904 "Municipal telephone and the tax-payers," 1 (4): 12.
1909 "A typical metropolitan telephone exchange," 1 (6): 1–5.
"To an operator," 1 (6): 11.
"Telephone service from an operator's standpoint," 1 (8): 10–11.
"Electric bells and buzzers," F.A. Field, 1 (8): 11.
1910 "Trans-atlantic service," 1 (11): 9.
"Operator and subscribers," 1 (11): 9a.
"Handling complaints," W.J. Stubbs, 1 (13): 1.
"Benefits of the farm telephone," 2 (2): 9.

No title, 2 (6): 1–2.
"An operator's influence," 2 (6): 11.
1911 "Responsibility of toll operators," P.L. Smith, 2 (9): 3–4.
"Dissatisfied subscriber," 2 (9): 4.
"Responsibility of tool operators," 2 (9): 4.
"Your best friend," 2 (10): 12.
"The telephone voice," 3 (1): 5.

TELEPHONY
1901 "The operating forces of modern telephone exchanges," H.S. Coyle, 2 (6): 218.
1903 "Modern telephone service," S.J. Larned, 5 (5): 300–02.
"Bell Company methods in Canada," 6 (3): 186–90.
"Rural telephone – Some amusing complications," 6 (6): 480–01.
1904 "The telephone mail", E.I. Innis, 7 (1): 71–3.
"The telephone Girl's voice," 7 (2): 126–27.
"The lady and the telephone," 7 (4): 309.
"The patient, facinating 'Hello Girl'", 7 (6): 418.
"The party line controversy," 7 (6): 452–3.
"The use of the telephone," 7 (6): 456–7.
"Weather forecast by telephone," M.C. Smythe, 8 (1): 12.
"Is the telephone making us left-eared?" P.S. Sissoni, 8 (1): 74–5.
"Treatment of telephone operators," 8 (2): 124–5.
"Bad telephone manners," 8 (2): 130.
"Heard on party line," 8 (3): 211.
"Entertainment as freedom," 8 (3): 232.
"Telephones and manners," 8 (4): 311.
1905 "The 'Hello' girl and 'Central,'" 9 (1): 38–42.
"Eliminating isolation from farm life," 9 (3): 255–8.
"Where 'Central' goes to school," 9 (3): 259–60.
"Soft voices ensnare men's hearts," 9 (4): 328.
"A study of the telephone girl," 9 (5): 388–90.
"Telephone has improved American voice," W.H. Kenney, 9 (5): 428–9.
"Lays telephonitis to women," 10 (3): 211.
"Trespass by telephone," 10 (3): 211.

"Shopping by night by telephone," 10 (3): 221b.
"Tragedies and romances of the telephone," M.E. Barry, 10 (4): 292–4.
"Telephone brakes the voice soft," 10 (5): 360.
"Troubles over the telephone," 10 (6): 429.
"A breach of the peace", 10 (6): 433.

Diocese of Montreal Archives

1843–1935   *Rapports Pastoraux*. Montreal.
1870–1925   *Mandements, Lettres Pastorales, Circulaires et autres documents*. Vols. 5 to 16.

Montreal Municipal Archives

1879–1905   Telephone directories, Montreal and Toronto.
1881–91   Montreal map, C.E. Goad #1–2, plate 50.
1942a   "Le développement de Montréal," *Le quartier latin*, 27 Mar.
1942b   "Montréal de 1867 à 1914", *La bonne parole*, April.
1947   "Montreal in 1880", *Le Devoir*, 22 Sept.

Other sources

Adler, R. "The Telephone and the Evolution of the American Metropolitan System." In I. de S. Pool, ed., *The Social Impact of the Telephone*, 318–41. Cambridge: MIT Press 1977.

"Action at a Distance." *Scientific American* 77 (1914): 39.

Acton, J., et al. *Women at Work: Ontario 1850–1930*. Toronto: The Women's Press 1974.

"A Fire Alarm System which Telephones its Message." *Scientific American* 122 (1920): 494.

Alcott, L.M. *Work*. New York: Schocken 1873.

Alzon, C. *La femme potiche et la femme bonniche*. Paris: Maspero 1979.

Ames, H.B. *The City Below the Hill*. Toronto: University of Toronto Press 1897.

Amstrong, C., and H.V. Nelles. *Monopoly Moment*. Philadelphia: Temple University Press 1986.

"An American Telephone Newspaper." *Literary Digest* 44 (1912): 528–9.

Andrew, S.L. "World's Telephone Statistics, 1921." *Bell Telephone Quarterly* 1, no. 3 (1922): 45–54.

– "World's Telephone Statistics, 1922." *Bell Telephone Quarterly* 2, no. 4 (1923): 255–6.

"A Pocket Telephone." *Literary Digest* 44 (1912): 639.

Aronson, S.H. "Bell's Electrical Toy: What's the use? The Sociology of Early Telephone Usage." In I. de S. Pool, ed., *The Social Impact of the Telephone,* 15–39. Cambridge: MIT Press 1977.

Attali, J., and Y. Stourdze. "The Birth of the Telephone and Economic Crisis: The Slow Death of the Monologue in French Society." In I. de S. Pool, ed. *The Social Impact of the Telephone,* 97–111. Cambridge: MIT Press 1977.

"A Vest-Pocket Transmitter for the Telephone." *Scientific American* 106 (1912): 112.

Babe, R.E. "Control of Telephones: The Canadian Experience." *Canadian Journal of Communication* 13, no. 2 (1988): 16–29.

"Back to the Land – and the Telephone." *Spectator* 96 (1906): 530–1.

Banks, J.A., and O. Banks. *Feminism and Family Planning in Victorian England.* New York: Schocken 1977.

Banning, W.P. "The Growing Use of Telephone 'Conference service.'" *Bell Telephone Quarterly* 14, no. 30 (1936) 146–56.

Barrett, R.T. "First Lady of the Switchboard." *Bell Telephone Quarterly* 14 (1935a): 105.

– "The Changing Years as Seen from the Switchboard." *Bell Telephone Quarterly* 14 (1935b): 285–8.

– "The Telephone as a Social Force." *Bell Telephone Quarterly* 19 (1940): 129–38.

Barrett, W.F. "The Electric Telephone: Its Earliest and Latest Development." *Goodwords* 19 (1878a): 277–84.

– "The Telephone, Its History and Its Recent Improvements." *Nature* 19 (1878b): 12–14; 56–9.

Bedford, H. "The Telephone." *The Irish Monthly,* part I, 7 (1879): 337–45; part II, 7 (1879): 406–13.

"Behind the Scene at 'Central.'" *Booklovers' Magazine* 2 (1903): 390–401.

Benston, M. "The Political Economy of Women's Liberation." *Monthly Review* 21 (1969): 13–27.

Bernard, E. *Long Distance Feeling.* Vancouver: New Books 1982.

Blauner, R. *Alienation and Freedom.* Chicago: University of Chicago Press 1964.

Bolitho, H. "The Telephone and the Biographer." *Spectator* 155 (1935): 90–1.

Bower, J.A. "The Photophone." *Goodwords* 22 (1881): 139–41.

Braverman, H. *Labor and Monopoly Capital.* New York: Monthly Review Press 1974.

Briggs, A. "The Pleasure Telephone: A Chapter in the Prehistory of the Media." In I. de S. Pool, ed., *The Social Impact of the Telephone,* 40–68. Cambridge: MIT Press 1977.

Briskin, L. "Domestic Labour: A Methodological Discussion." In B. Fox, ed., *Hidden in the Household,* 135–72. Toronto: The Women's Press 1980.

Brooks, J. *Telephone*. New York: Harper and Row 1974.

– "The First and Only Century of Telephone Literature." In I. de S. pool, ed., *The Social Impact of the Telephone*, 208–24. Cambridge: MIT Press 1977.

Bruce, R.V. *Bell: Alexander Graham Bell and the Conquest of Solitude*. Boston: Little, Brown & Co. 1973.

Burawoy, M. *Manufacturing Consent*. Chicago: University of Chicago Press 1979.

Burlingame, R. *Engines of Democracy*. New York: Charles Scribner 1940.

Burton, C.L. *A Sense of Urgengy*. Toronto: Clarke, Irwin & Co. 1952.

Calhoun, A.W. *A Social History of the American Family from 1865 to 1919*. Vol. 3. New York: Barnes and Noble 1919.

Carey, J. "Telegraph to Computers: Technology and History." Paper presented at The International Colloquium on Communication, University of Toronto 1985.

Carty, J.J. "The Telephone Development." *Bell Telephone Quarterly* 1, no. 1 (1922a): 1–11.

– "Ideals of the Telephone Service." *Bell Telephone Quarterly* 1, no. 3 (1922b): 23–37.

– "Semi-Centennial of the Telephone." *Bell Telephone Quarterly* 5, no. 1 (1926a): 1–11.

– "Episodes in Early Telephone History." *Bell Telephone Quarterly* 5, no. 2 (1926b): 59–70.

Casson, H.N. *The History of the Telephone*. Freeport: Books for Libraries Press 1910a.

– "The Telephone as It Is Today." *World's Work* 19 (1910b): 127–75.

– "The Future of the Telephone." *World's Work* May (1910c): 1908–13.

– "The Social Value of the Telephone." *Independent* 71 (1911): 899–906.

Chamberlain, M. *Fen Women: A Portrait of Women in an English Village*. London: Routledge and Kegan Paul 1975.

Cherry, C. "The Telephone System: Creator of Mobility and Social Change." In I. de S. Pool, ed., *The Social Impact of the Telephone*, 112–26. Cambridge: MIT Press 1977.

Clark, C.S. *Of Toronto the Good*. Montreal: The Toronto Publishing Co. 1898.

Cockburn, C. "The Relations of Technology: What Implications for Theories of Sex and Class." In R. Crompton and M. Mann eds., *Gender and Stratification*, 74–85. Cambridge, UK: Polity 1986.

Collectif Cleo. *L'histoire des femmes du Québec*. Montreal: Quinze 1985.

Collins, R.J. *A Voice from Afar*. Toronto: McGraw-Hill Ryerson 1977.

Conseil du statut de la femme, ed. *Sortir la maternité du laboratoire*. Quebec City: Gouvernement du Québec 1988.

Copp, T. *The Anatomy of Poverty: The Condition of the Working Class in Montreal 1897–1929*. Toronto: McClelland and Stewart 1974.

Corrigan, P., and P. Willis. "Cultural Forms and Class Mediations." *Media, Culture and Society* 2 (1980): 297–312.

Côté, Y., "La téléphonie au Québec." *Communication Information* 2, no. 1 (1977): 57–66.

Cross, D.S. "La majorité oubliée: le rôle de femmes à Montréal au 19ième siècle." In M. Lavigne and Y. Pinard, eds. *Les femmes dans la société québécoise*, 33–59. Montreal: Boréal Express 1977.

"Curiosities of the Telephone." *Chambers's Journal* 60 (1883): 17–20.

Curtis, B. "Capital, the State and the Origins of the Working-Class Household." In B. Fox, ed., *Hidden in the Household*, 101–34. Toronto: The Women's Press 1980.

Danylewycz, M., and A. Prentice. "Teachers' Work." *Labour/Le Travail* 17 (1986): 58–80.

Danylewycz, M., et al. "The Evolution of the Sexual Division of Labour in Teaching: A Nineteenth Century Ontario and Quebec Case Study." *Social History* 16, no. 31 (1983): 81–109.

DeBonville, J. *Jean-Baptiste Gagnepetit*. Montreal: Aurore 1975.

De la Haye, Y., ed. *Marx and Engels on the Means of Communication*. New York: International General 1980.

Dilts, M.M. *The Telephone in a Changing World*. New York: Longman's Green 1941.

Dixon, A.F. "Development of Communication Systems." *Bell Telephone Quarterly* 2, no. 1 (1926): 67–8.

Donald, R. "The State and the Telephone: A story of a Betrayal of Public Interests." *The Contemporary Review* 74 (1898): 530–46.

"Edison's Telephone." *Nature* 19 (1878): 471–2.

Edwards, R. *Contested Terrain*. New York: Basic Books 1979.

Ellul, J. *The Technological Society*. New York: Vintage 1964.

Ewen, S., and E. Ewen. *Channels of Desire: Mass Images and the Shaping of American Consciousness*. New York: McGraw-Hill 1982.

Fahmy-Eid, N., and M. Dumont. *Maîtresses d'école, maîtresse de maison*. Montreal: Boreal Express 1983.

Fetherstonhough, R.C. *Charles Fleetford Sise, 1834–1918*. Montreal: Gazette 1944.

Finlay-Pelinski, M. "Pour une épistémologie de la communication: au-delà de la représentation et vers la pratique." *Communication Information* 5, nos 2–3 (1983): 5–34.

Fischer, C. "The Revolution in Rural Telephony, 1900–1920." *Journal of Social History* 21, no. 1 (1988a): 5–26.

– "'Touch Someone': The Telephone Industry Discovers Sociability." *Technology and Culture* January (1988b): 32–61.

– "Gender and the Residential Telephone in North America, 1890–1940: Technologies of Sociability." *Sociological Forum*, in press.

Flynt, J. "Telephone and Telegraph Companies as Allies of Criminal Pool Rooms." *Cosmopolitan* 43 (1907): 50–7.

Fortner, R.S. "Communication and Regional/Provincial Imperatives." *Canadian Journal of Communication* 6, no. 4 (1980): 32–46.

Foucault, M. "Questions of Method." *Ideology and Consciousness* 80 (1981): 3–14.

– "Subject of Power." *Critical Inquiry* 8 (1982): 777–95.

"Future of Telegraphs and Telephones, The." *Chambers's Journal* 62 (1885): 439–40.

Gaskell, Mrs. *Wives and Daughters*. New York: Everyman's 1866.

Gherardi, B. "The Bell System." *Bell Telephone Quarterly* 4, no. 4 (1925): 225–65.

Goody, J., and I. Watt. "The Consequence of Literacy." *Comparative Studies in Society and History* 5 (1963): 304–45.

Grumet, M. "Pedagogy for Patriarchy: The Feminisation of Teaching." *Interchange* 12 (1981): 165–84.

Haller, J.S. and R.M. Haller. *The Physician and Sexuality in Victorian America*. New York: Norton 1974.

Harvey, D. *The Urbanization of Capital*. London: Basil Blackwell 1985.

– "Flexible Accumulation through Urbanization: Reflections on 'Post-Modernism' in the American City." *Antipode* 19, no. 3 (1987): 260–86.

Hastie, A.H. "The Telephone Tangle and the Way to Untie It." *The Fortnightly Review* 70 (1898): 894.

Hayden, D. *The Grand Domestic Revolution*. Cambridge: MIT Press 1981.

Helmle, W.C. "The Relation between Telephone Development and Growth and General Economic Conditions." *Bell Telephone Quarterly* 4, no. 1 (1925): 8–21.

Heron, C., and B.D. Palmer. "Through the Prism of the Strike: Industrial Conjunct in Southern Ontario, 1901–14." *Canadian Historical Review* 58 (1977): 423–58.

Heyer, G.K. "The Telephone and the Railroad." *Scientific American Supplement* 70 (1910): 388–9.

Honoré, F. "The Secret Telephone." *Scientific American* 121 (1919): 555.

Houston, D.F. "The Young Man and Modern Business." *Bell Telephone Quarterly* 6, no. 2 (1927): 77–89.

Husband, J. "Telephone." *Atlantic Monthly* 114 (1914): 330–1.

Innis, H.A. *The Bias of Communication*. Toronto: University of Toronto Press 1968.

– *Empire and Communications*. Toronto: University of Toronto Press 1972.

"Introduction of the Telephone in Great Britain." *Westminster Review* 109 (1878): 208–21.

Jewett, F.B. "Electrical Communication: Past, Present, Future." *Bell Telephone Quarterly* 14, no. 3 (1935): 167–200.

– "The Social Implication of Scientific Research in Electrical Communication."
Bell Telephone Quarterly 15, no. 4 (1936): 205–18.

Katz, M. *The People of Hamilton, Canada West: Family and Class in a Mid-Nineteenth Century City.* Cambridge: Harvard University Press 1975.

Kealy, L., ed. *Women at Work, Ontario 1850–1930.* Toronto: Women's Press 1974.

Keller, S. "The Telephone in New (and Old) Communities." In I. de S. Pool, ed., *The Social Impact of the Telephone,* 281–98. Cambridge: MIT Press 1977.

Kelly, A.M. *Three Victorian Telephone Directories.* New York: A.M. Kelly 1970.

King, W.L.M. *Industry and Humanity.* Toronto: Thos. Allen 1918.

Kingsbury, J.E. *The Telephone and Telephone Exchanges.* New York: Longman's, Green & Co. 1915.

Kramarae, C. *Technology and Women's Voices.* New York: Routledge and Kegan Paul 1988.

Kuyek, J.N. *The Phone Book: Working at the Bell.* Kitchener: Between the Lines 1979.

"La Famille." *Le monde illustré* 1884.

Lalonde, M. "La femme de 1837–38: complice ou contre révolutionnaire?" *Liberté* 7, nos 1–2 (1965): 146–73.

Lang, A. "Telephones and Letter-writing." *Critic* 48 (1906): 507–8.

Langdon, W.C. "Founders of Bell System." *Bell Telephone Quarterly* 2, no. 4 (1923): 266–80.

Langton, J. *Natural Enemy.* New York: Penguin 1987.

Latham, R.F. "The Telephone and Social Change." In B.D. Singer, ed., *Communication in Canadian Society,* 19–36. Vancouver: C. Clark 1975.

Lavigne, M., and Y. Pinard. *Les femmes dans la société québécoise.* Montreal: Boréal Express 1977.

Lavigne, M., and J. Stoddart. "Ouvrières et travailleuses montréalaises, 1900–1940." In M. Lavigne and Y. Pinard, eds. *Les femmes dans la société québécoise,* 125–43. Montreal: Boréal Express 1977.

Lecompte, P.E. *Sir Joseph Dubuc.* Montreal: Messager 1923.

Linteau, P.A., A. Durocher, and J.C. Robert. *Quebec a History, 1867–1929.* Toronto: Lorimer 1983.

Lockwood, T.D. *Practical Information for Telephonists.* New York: W.J. Johnstone 1893.

"London-Paris Telephone, The." *Nature* 14 (1876): 30–2.

Lowe, G.S. "Women, Work and the Office: The Feminization of Clerical Occupations in Canada, 1901–1931." *The Canadian Journal of Sociology* 5, no. 4 (1980): 361–81.

Lukacs, G. *Histoire et conscience de classe.* Paris: Minuit 1960.

Luxton, M. *More than a Labour of Love.* Toronto: The Women's Press 1980.

Lyon, O.C. "The Telephone Directory." *Bell Telephone Quarterly* 2, no. 3 (1924a): 187–91.

– "The Telephone Directory." *Bell Telephone Quarterly* 3, no. 3 (1924b): 175–85.

– "The Telephone Directory." *Bell Telephone Quarterly* 3, no. 3 (1924c): 187–91.

– "The Telephone Directory." *Bell Telephone Quarterly* (1927): 187–91.

– "Answering the Question 'Number Please.'" *Bell Telephone Quarterly* 7, no. 2 (1928): 95–106.

Maddox, B. "Good Jobs for Girls." *Telecommunication Journal* (1975, Dec.): 711.

– "Women and the Switchboard." In I. de S. Pool, ed. *The Social Impact of the Telephone*, 262–80. Cambridge: MIT Press 1977.

Mann-Trofimenkoff, S. "Henri Bourassa et la question des femmes." In M. Lavigne and Y. Pinard, eds., *Les femmes dans la société québécoise*, 108–24. Montreal: Boréal Express 1977.

Martin, M. "Communication and Social Forms: A Study of the Department of the Telephone System, 1876–1920." Unpublished Ph. D. thesis, Department of Sociology, University of Toronto, Toronto 1987.

– "Feminisation of the Labour Process in the Communication Industry." *Labour/Le travail* 22 (1988, fall): 139–62.

– "Rulers of the Wires? Women's Contribution to the Structure of Means of Communication." *Journal of Communication Inquiry* 12, no. 2 (1988): 89–103.

– "Capitalizing on the 'Feminine' Voice." *Canadian Journal of Communication* 14, no. 3 (1989): 42–62.

– "Communication and Social Forms: The Development of the Telephone, 1876–1920." *Antipode: A Radical Journal of Geography*. In press.

Marvin, C. "Space, Time, and Captive Communications." In M.S. Mander, ed. *Communications in Transitions*, 20–38. New York: Praeger 1983.

– *When Old Technologies were New*. Cambridge: Oxford 1988.

Marx, K. *Le capital*. Vol. 1. Montreal: Nouvelle Frontière 1867.

– *A Contribution to the Critique of Political Economy*. New York: International 1959.

– *Grundrisse*. New York: Vintage 1973.

Marx, K., and F. Engels. *Le manifeste du parti communiste*. Paris: Éditions Sociales 1850.

– *The German Ideology*. Moscow: Progress 1932.

Mattelart, A. "For a Class Analysis of Communication." In A. Mattelart and S. Siegelaub, eds. *Communication and Class Struggle: Capitalism and Imperialism*, 17–67. New York: International General 1979.

– "For a Class and Group Analysis of Popular Communication Practices." In A. Mattelart and S. Siegelaub, eds., *Communication and Class Struggle: Liberation and Socialism*, 23–70. New York: International General 1983.

Mattelart, M. "The Feminine Version of the Coup d'État." In J. Nash and H. Icken Safa, eds. *Sex and Class in Latin America*, 279–301. New York: Praeger 1976.

— "Notes on 'Modernity': A Way of Reading Women's Magazines." In A. Mattelart and S. Siegelaub, eds., *Communication and Class Struggle: Capitalism and Imperialism*, 158–70. New York: International General 1983.

— "Can Industrial Culture Be a Culture of Difference?" *Border/Lines* 3 (1985a): 18–22.

— Interview. *Border/Lines* 3 (1985b): 19–21.

— *Women, Media and Crisis*. London: Comedia 1986.

Mayer, M. "The Telephone and the Use of Time." In I. de S. pool, ed., *The Social Impact of the Telephone*, 225–45. Cambridge: MIT Press 1977.

McComber, J.E. *Mémoires d'un bourgeois de Montréal, 1874–1949*. Montreal: Hurtubise 1980.

McEachen, S. "A Feminist Analysis of Raymond Williams' Theories of Culture." Unpublished paper, OISE, Toronto 1985.

McLuhan, M. *Understanding Media*. New York: Signet 1964.

Miner, H. *St. Denis: A French Canadian Parish*. Chicago: University of Chicago Press 1939.

"Mr Fawcett and the Telephone." *Spectator* 57 (1884): 1167.

Munro, J. "On the Telephone, an Instrument for Transmitting Musical Notes by Means of Electricity." *Nature* 14 (1876): 30–2.

— "Electricity and its Uses." *Leisure Hour* (1882): 309–12.

"Notes." *Nature* 19 (1878): 46.

"Notes." *Nature* 21 (1879): 115.

"Notes." *Nature* 23 (1880): 63.

"Notes." *Nature* 23 (1881): 590.

"Notes." *Nature* 26 (1882): 66, 257.

"Notes." *Nature* 58 (1898): 39, 524.

Ogle, E.B. *Long-Distance Please*. Toronto: Collins 1979.

Ollman, B. *Alienation: Marx's Concept of Man in Capitalist Society*. Cambridge: Cambridge University Press 1983.

"On Mulock committee." Toronto *Globe*, 31 March 1905.

Patten, W. *Pioneering the Telephone in Canada*. Montreal: Pioneers' 1926.

Perrine, J.O. "The Development of the Transmission Circuit in Communication." *Bell Telephone Quarterly* 4, no. 2 (1925): 114–31.

Pike, R. "From Luxury to Necessity and Back Again? Canadian Consumers and the Pricing of Telephone Services in Historical and Comparative Perspective." Paper presented at the 1985 Annual Conference of the international Communication Association, Honolulu, Hawaii, May 1985.

— "Kingston Adopts the Telephone." *Urban History Review* 18, no. 1 (1989): 32–47.

Pike, R., and V. Mosco. "Canadian Consumers and Telephone Pricing." *Telecommunications Policy* (1986, March): 17–32.

Pool, I. de S., ed. *The Social Impact of the Telephone.* Cambridge: MIT Press 1977.

– "Extended Speech and Sounds." In R. Williams, ed., *Contact: Human Communication and its History,* 169–82. London: Thames and Hudson 1981.

– *Forecasting the Telephone.* Norwood: Ablex 1983.

"Post Office and the Telephone, The." *Spectator* 57 (1884): 1068–9.

Pound, A. *The Telephone Idea: 50 Years After.* New York: Greenberg 1926.

Preece, W.H. "Recent Progress in Telephony." *Nature* 26 (1882): 516–19.

Prentice, A. "The Feminization of Teaching in British North America and Canada, 1845–1875." *Social History* 3, no. 15 (1975): 5–20.

"Problem of the Telephone, The." *Scientific American* 48 (1883): 96.

Probyn, E. "Travels in the Postmodern: Making Sense of the Local." In L. Nicholson, ed., *The Politics of Method.* New York: Routledge 1989.

Rakow, L.F. "Rethinking Gender Research in Communication." *Journal of Communication* 36, no. 4 (1986): 11–26.

"Recording Telephones." *Nature* 62 (1900): 371–3.

Rhodes, F.L. "Development of Cables Used in the Bell System." *Bell Telephone Quarterly* 2 no. 2 (1923a): 94–106.

– "How Telephone Wires Were First Put Underground." *Bell Telephone Quarterly* 2, no. 4 (1923b): 240–54.

– "Cable Development and Research." *Bell Telephone Quarterly* 3, no. 1 (1924): 30–40.

– "Outwitting the Weather." *Bell Telephone Quarterly* 6, no 1 (1927): 21–31.

– *Beginning of Telephony.* New York: Arno 1929.

Richardson, B.W. "Tricycling in Relation to Health." *Goodwords* 23 (1882): 734–7.

Richter, F.E. "The Telephone as an Economic Instrument." *Bell Telephone Quarterly* 4, no. 4 (1925): 281–95.

Rothschild, J., ed. *Machine ex Dea.* New York: Pergamon 1983.

Ryan, M.P. "The Power of Women's Networks." In J.L. Newton, M.P. Ryan, and J.R. Walkowitz, eds. *Sex and Class in Women's History,* 167–86. London: History Workshop Series 1983.

Samuel, R., ed. *Village Life and Labour.* London: Routledge & Kegan Paul 1975.

Sangster, J. "The 1907 Bell Telephone Strike: Organizing Workers." *Labour/Le Travail* 3 no. 3 (1978): 109–30.

Sayers, D.L. *The Unpleasantness at the Bellona Club.* London: NEL 1921.

Siegelaub, S. "A Communication on Communication." In A. Mattelart and S. Siegelaub, eds. *Communication and Class Struggle: Capitalism and Imperialism,* 11–21. New York: International General 1979.

– "Working Notes on Social Relations in Communication and Culture." In A. Mattelart and S. Siegelaub, eds. *Communication and Class Struggle: Liberation and Socialism,* 11–16. New York: International General 1983.

"Simultaneous Telegraphing and Telephoning." *Scientific American* 90 (1904): 459.

Singer, D.B. *Social Functions of the Telephone.* Palo Alto: R & E. Research 1981.

Smythe, D.W. *Dependency Road.* Norwood: Ablex 1981.

Spofford, H.P. "A Rural Telephone." *Harper's Magazine* 118 (1909): 830–7.

Stanton, E.C. *Eighty Years and More: Reminiscences 1815–1897.* New York: Schocken 1897.

Stanton, L.W. "The Telephone System of the Future." *Scientific American Supplement* 61 (1906): 25, 399.

Strasser, S. *Neverdone.* New York: Pantheon 1982.

"Telephone, The." *Bloor Watchman,* 30 May 1929.

"Telephone, The." *Chambers's Journal* 76 (1899): 310–13.

"Telephone, The." *Nature* 16 (1877): 403–4.

"Telephone and Railways, The." *Scientific American* 70 (1910): 388.

"Telephone and the Doctor, The." *Literary Digest* 44 (1912): 639.

"Telephone Exchange, The." *Spectator* 52 (1879): 1187–8.

"Telephone Exchange in the United States, The." *Nature* 21 (1879): 495–7.

Thayer, H.B. "Evolution – Not Revolution." *Bell Telephone Quarterly* 3, no. 3 (1924): 133–9.

– "A Decade in Retrospect: 1914–1924." *Bell Telephone Quarterly* 4, no. 1 (1925): 1.

Thompson, E.P. *The Making of the English Working Class.* Harmondsworth: Penguin 1963.

Thompson, S.P. *Philipp Reis: Inventor of the Telephone.* London: E. & F.N. Spon 1883.

Tonnies, F. *Community and Society.* New York: Harper and Row 1963.

"To Stop Telephone-Eavesdropping." *Literary Digest* 49 (1914): 733.

Vail, T.N. "Public Utilities and Public Policy." *Atlantic Monthly* 111 (1913): 307–19.

Van Deventer, H.R. "A Telephone Transmitter without a Mouth Piece." *Scientific American* 108 (1913): 468.

Walkerdine, V. "On the Regulation of Speaking and Silence: Subjectivity, Class and Gender in Contemporary Schooling." In C. Steedman, C. Urwin, and V. Walkerdine, eds. *Language, Gender and Childhood,* 203–41. London: Routledge & Kegan Paul 1985.

Waterson, K.W. "Service in the Making." *Bell Telephone Quarterly* 1, no. 3 (1922): 26–33.

Watson, J.C. "The Telephone." *Nature* 19 (1878): 95–6.

Wells, G.H. *Anticipation of the Reaction of Mechanical and Scientific Progress upon Human Life and Thought.* New York: Harper and Brothers 1902.

Wharton, E. *The House of Mirth.* New York: Berkley 1905.

- *Summer.* New York: Berkley 1917.

- *The Age of Innocence.* New York: Scribner's 1921.

Whicker, M.L., and J.J. Kronenfeld. *Sex Role Changes: Technology, Politics, and Policies.* New York: Praeger 1985.

Williams, R. *Contact: Human Communication and its History.* London: Thames and Hudson 1981a.

- *Culture.* Glasgow: Fontana 1981b.

- *Problem in Materialism and Culture.* London: Verso 1982.

Willis, P., and P. Corrigan. "Orders of Experience: The Differences of Working Class Cultural Forms." *Social Text* 7 (1983): 85–103.

Woolf, V. *A Room of One's Own.* Toronto: Granada 1929.

Index

Activities, 10, 12, 45
Analysis: feminist, 5, 7, 51,
  52, 80, 161, 172; political-economic, 5-7, 51,
  80, 172
Armstrong, C., and Nelles,
  H.V., 172

Bell, A.G., 29, 36, 57, 58,
  137
Bell Telephone Co.: 9, 11,
  15, 19–48, 50, 55, 56,
  57, 60, 67, 68, 73, 75,
  79, 80, 91, 101, 103,
  105–7, 110, 118, 123,
  131, 132, 134, 136,
  137, 141–6, 149, 155;
  Act of Incorporation, 29,
  30, 39, 44, 46, 60; competition, 29, 30, 33, 40,
  41, 48; internal policies,
  92, 104, 105, 141; man-
  agers, 29, 31, 33, 36, 38,
  40, 47, 48, 50, 51, 54,
  55, 57–9, 63–5, 71, 74,
  76, 80, 91, 92, 101,
  103–5, 108, 116, 131,
  135, 146, 149, 152;
  monopoly, 12, 14, 29,
  32, 33, 36, 40–9, 52,
  108, 167; moral control,
  52, 64, 66, 67, 74, 94,
  100; promotion, 36–8;
  revenues, 32, 33, 43, 77;
  rules and regulations, 62,

64–75, 92, 93, 103, 104,
  107
Business: men, 9, 14, 26,
  28, 29, 32, 35, 42, 44,
  56, 65, 123, 125, 136,
  140, 143, 146, 169; telephone, 11, 22, 26, 28,
  32, 39, 43, 45, 60, 93,
  96, 99, 100, 105, 115,
  121, 125, 136, 142, 148;
  transactions, 14, 24, 49,
  108, 134, 137, 143, 155,
  163, 169

Capital, 9, 16, 27, 28, 38,
  40, 43, 44, 108, 110,
  136, 168, 169
Capitalist: control, 27–9,
  49, 121, 127, 149, 150;
  developers, 28, 151;
  industries, 27, 49, 72,
  108, 112, 113, 115, 116,
  121; interest, 4, 38, 47,
  126, 127, 168, 169;
  monopoly, 16, 17, 52;
  production, 27, 39, 51,
  61, 108; society, 8, 17,
  28, 38, 52, 159, 163,
  172; system, 17, 40, 51
Capitalists, 9, 28, 30, 140,
  148, 167
Casson, H.N., 160, 163
Class difference, 10, 16,
  17, 94, 99, 107, 108,
  112–15, 121, 126, 127,

139, 141, 142, 144, 163,
  167
Class distribution, 111–13
Classes: ruling, 7, 9, 10,
  16, 17, 31, 35, 37, 38,
  53, 54, 57, 65, 91, 94,
  95, 99, 106, 107, 110–
  18, 121–9, 133, 134,
  140–6, 149–51, 154,
  156, 159, 160, 163, 170;
  rural, 31, 45, 47, 95,
  106, 128, 140; social, 5–
  7, 9, 16, 17, 28, 91, 92,
  94, 111, 141; working,
  31, 32, 34, 38, 41–3, 49,
  53, 57, 78, 91, 94, 95,
  101, 107, 110–18, 121,
  126, 127, 140, 144, 146,
  161, 163
Cockburn, C., 3, 154
Communication: accessibility of, 6, 16, 17, 38;
  forms of, 4, 5, 7, 8, 10,
  12, 16, 92, 110, 123,
  130, 136, 141–3, 146,
  154, 162; interactive, 11,
  14, 17, 22, 38, 51, 142;
  long-distance, 9, 10, 15,
  18, 24, 51; means of, 3,
  6–11, 15–17, 26, 28, 49,
  92, 108, 118, 126, 129,
  135, 137, 138, 141, 151,
  152, 155, 160, 163, 169,
  170, 172; process of, 6,
  8, 16; production of, 6,

96, 97, 108; systems of, 6, 7, 16, 17, 19, 25, 32, 72, 125, 126, 136; telephone, 11, 21, 22, 25, 27, 34, 36, 37, 44, 97–9, 124; types of, 6, 7, 16, 17, 109, 142, 158, 167
Curtis, B., 51

Dagger, F., 45, 46

Foucault, M., 61, 74, 80, 108

Gender: differentiation, 3, 7, 8, 10, 17, 167, 171; -oriented practices, 68, 126, 141–4; particularities, 5, 10

Haller, J.S., and Haller, R.M., 151
Harvey, D., 169
Heron, C., and Palmer, B.D., 77

Labour force, 57, 58, 65, 72, 74, 92, 113; control over, 57, 61, 65–7, 72, 79
Labour market, 51, 57, 60
Labour process, 12, 54, 61, 62, 65, 66, 69, 72, 75, 92, 97, 101, 107, 108, 131, 133, 172; feminization of, 50–2, 57–61, 80, 81; hierarchy in, 68–70; job characteristics in, 51, 55–8; transformation of, 12, 61
Lang, A., 160, 161
Legislatures, provincial, 24, 46, 47
Lukacs, G., 73

McFarlane, L.B., 20, 45, 48, 55, 58, 68, 73, 74, 137, 138, 144
McLuhan, M., 15, 152
Marvin, C., 6, 110, 140,

146
Mattelart, A., 6
Mattelart, M., 5, 8, 141, 149, 169, 170
Men's: activities, 128, 141, 158; characteristics, 52–4, 56; control, 5, 53, 146; domination, 5, 53; monopoly, 3, 154; subordination, 53; world, 38
Montreal, 24, 28–30, 33–5, 40–2, 50, 55, 56, 72, 73, 76, 77, 107, 111–18, 121, 123–31, 136
Mullock Committee, 33, 45, 47, 48
Municipal Councils, 23, 24, 40, 46, 105
Municipalities, 23, 41, 45–7

Newspapers: advertisements, 28, 35, 36, 38, 46, 123, 130, 140, 146–51, 155, 160; articles, 36, 91, 95, 99, 103, 104, 106, 130, 133, 144, 147, 150, 151
Nielson, H., 33

Ontario, 24, 26, 28, 33, 35, 39, 44–6, 54, 56, 57, 59, 70, 100, 107, 132, 136, 137, 142, 143, 145
Ontario Railway and Municipal Board, 101
Ontario Select Committee, 45, 145
Operators: characteristics of, 56, 60, 63, 92, 93; chief, 64, 65, 66, 68–70, 72, 92; company rules for, 62, 64–6, 92, 93; contribution of, 12, 52, 62, 91, 97, 109; control over, 61–72, 74, 75, 78, 80, 92–5, 105; devil, 69, 74, 75, 97, 102, 104, 106, 107; discipline, 61, 62, 65, 69, 70, 72–5, 93,

104, 107; exploitation, 63, 65, 67, 75, 77, 78, 81, 101, 172; female, 4, 11, 12, 25, 36, 43, 50–2, 54, 56, 57, 61–9, 76, 77, 79, 80, 91–3, 95–102, 107, 108, 110, 168, 169; hiring conditions, 51, 55, 57–9, 78, 93; labour force, 50, 51, 54, 56, 58–60, 65, 66, 72, 74, 108; living expenses, 76, 78; male, 50, 52, 55–8, 81; occupation of, 4, 11, 50–2, 67, 168, 169; perfect, 59, 60, 61, 65, 66, 73, 75, 95, 102–4, 106, 107; process of feminization of the occupation of, 12, 51, 52, 57–61, 80, 81; production, 11, 12, 21, 50, 56, 60–9, 71–5, 93–5, 97, 101, 108; skills, 56–8, 108; strike, 54, 76–9; subjectification, 67, 72, 73; subjection, 4, 12, 50, 51, 54, 55, 57, 65–70, 72, 74, 81, 92, 93, 95, 97, 104, 108; supervision, 67, 72, 73; training, 51, 52, 55, 58, 61, 65–70, 73–5, 91–5, 101; unionization, 79, 80; wages, 51, 55, 56, 59, 60, 61, 63, 64, 71, 76–9; work, 51, 52, 55, 56, 60, 62, 63, 65, 67–70, 73, 92, 93, 95, 97–100, 102–8, 123, 138, 139, 148, 153; working conditions, 12, 19, 60, 62–4, 66–8, 70–2, 74, 76–80, 92, 93, 97, 101, 147
Ottawa, 41, 46–8, 59, 103, 136, 137

Practices: class-oriented, 6, 96, 127, 139, 141, 145; cultural, 4, 6, 10, 13, 16,

17, 97, 126, 141, 155, 158–61, 163, 167, 169, 170, 172; group telephone, 10, 48, 109, 133, 134, 141; social, 5–7, 10, 26, 123, 128, 129, 141, 142, 145, 155, 158, 159, 161, 169, 170, 172; telephone, 8–10, 12, 129, 133, 134, 141–3, 146, 149–53, 161, 167, 170

Privacy and secrecy, 18, 20–2, 24, 26, 27, 32, 39, 44, 49, 50, 52, 67, 108, 128, 131, 142–5, 151, 152, 158; devices, 19–22; lack of, 20, 45, 123, 124, 144–6, 162; for operators, 64, 67, 69, 107, 144; for subscribers, 65

Profit, 14, 30, 31, 40, 43, 48, 65, 72, 77, 79

Profitable development, 11, 14, 30, 31, 33, 36, 65, 67, 69, 81, 131, 133, 137, 139, 141, 167, 172

Provincial: Enabling Acts, 23, 44, 107, 145; legislatures, 24, 46, 47; regulation, 14, 28, 39, 40, 43, 44

Quebec, province of, 26, 28, 33, 35, 44, 47, 107, 145

Quebec City, 23, 33, 35, 136

Radcliffe, J., 19

Railway Commission, 30, 40, 44

Resistance: of operators against companies, 10–12, 29, 30, 40, 41, 43–5, 47–9, 61, 62, 65, 67, 68, 75–81, 99, 104, 106–8, 144, 145, 167, 169, 171, 172; of operators against

subscribers, 92, 93, 95–109; of subscribers, 146, 147, 169, 171

Ryan, M.P., 128, 168

Sangster, J., 54, 77

Siegelaub, S., 6, 11

Sise, C.F., 29, 33, 36, 45–7, 56, 57, 63, 80, 94, 132, 134, 148

Smythe, D.W., 45

Social: attitudes, 51–4; control, 5, 17; differentiation, 16–17; groups, 9, 10–12, 16, 17, 26, 28, 34, 38, 40, 44, 49, 140, 143, 163, 167; order, 16, 37, 38; relations, 4, 14; status, 51, 52, 60, 61, 78, 127, 168

Society: capitalist, 5, 17, 38, 167, 169; patriarchal, 5, 52, 168; late-Victorian, 9, 52, 53, 59, 62, 110, 123, 126, 134, 142, 146, 151, 155, 172

State: agencies, 23, 43, 45, 47, 115; Federal Enabling Acts, 24, 39, 40, 44, 107, 142, 158; Federal Railway Commission, 39, 40, 44; interventions, 9, 14, 32, 39, 44, 46, 48, 115, 145, 158, 167, 169, 172

Strasser, S., 147

Subscribers: 4, 6, 9, 10, 12, 14, 18–23, 25–8, 30–42, 48–50, 52, 55–8, 60, 61, 63–9, 73, 75, 81, 91–3, 96–103, 106, 110, 115, 118, 123, 124, 129, 131, 132, 134, 136, 140, 142–4, 146; complaints, 56, 73, 99, 104–6, 131; disciplining, 134, 135, 145; education of, 91, 96, 99–101, 108, 129–35, 157, 159, 162; pressures from, 40, 41

Swinyard, T., 33, 144

Technological: advance, 24, 25, 32, 58, 60, 65, 108; characteristics, 11, 15, 19, 63, 121, 151, 153, 158, 165; instrument, 25, 52, 73, 93, 96, 98, 105, 108, 110, 141; knowledge, 55, 56; uses, 15, 16, 96

Technology: control of, 9, 15, 16, 21, 32; development of, 3, 4, 10, 11, 14, 15, 18, 21, 26, 27, 28, 32, 33, 49, 130, 141; effect of, 4, 96, 158, 159; technology, 10–12, 14–17, 21, 31, 63, 97, 100, 110, 123, 124, 130, 134, 135, 140–3, 149–51, 154, 161, 162, 164, 167, 169, 171; uses of, 3–6, 10, 11, 14, 16

Telegraph: messengers, 50, 52, 54, 56, 107; poles and wires, 23, 24; rates, 43; technology, 15, 28, 36, 118, 124, 125, 126, 129, 130, 136

Telegraphy: 15, 22, 155, 160

Telephone: accessibility, 11, 14, 17, 26, 28, 32, 33, 38, 42, 110, 115, 121, 124–6, 135, 139; availability, 17, 168; business, 11, 28, 29, 52, 63, 69, 72, 80, 94, 108, 135, 148, 155; cables and wires, 18, 22–5, 75, 135, 143, 145, 146; calls, 18, 21, 22, 24, 25, 31, 44, 52, 56, 61, 65–9, 73, 74, 97, 108, 116, 133, 141, 143, 148, 150, 151, 154, 159, 160, 162; central, 25, 26, 123, 136, 138, 143; collectivity, 28, 136, 152, 154, 168; communication, 3, 22, 24, 25, 28, 45, 68, 149, 152; competition, 29, 30, 33, 39, 40, 41, 48; connections, 25, 32, 34, 39, 40, 41, 51, 100, 130, 134; control, 17, 22, 26, 28, 29; conversation, 20, 21, 24, 25, 37, 44, 69, 92, 115, 131, 133, 134, 142–4, 152, 153, 158, 162, 164; culture of the, 4, 12, 139, 142, 155–63, 171; development, 3, 4, 6, 7, 9–12, 14, 15, 17–23, 25–9, 32, 35, 38–40, 43, 46, 50, 52–4, 56, 60, 91, 92, 94, 97, 100, 102, 108, 109, 110, 112, 115, 121, 126, 129, 130, 139, 140–3, 145, 150, 155, 165, 167, 172; directory, 57, 100, 118, 130, 132, 148; ear, 161, 162; etiquette, 91, 101, 131, 132, 134, 145, 149, 151, 159, 160, 163; exchanges, 9, 25, 26, 28, 30–2, 36, 48, 52, 55, 56, 59, 62–4, 70, 72, 95, 97, 103, 111, 115, 116, 118, 121, 123, 124, 135–8, 140, 143, 144, 146, 152, 156; expansion, 4, 10, 44, 45, 48; eavesdropping, 21, 142, 144, 145, 152, 153; independent companies, 5, 9, 10, 12, 15, 17, 21, 27–30, 35, 40, 41, 47, 48, 50, 52, 54, 56, 57, 63, 64, 71, 91, 92, 94, 96, 97, 99, 101, 104, 108, 115, 124, 129, 133, 135, 136, 139, 140, 142, 144, 145, 147, 149, 150, 152–4, 159; industry, 14, 22, 23, 26, 27, 47, 50–2, 54, 56, 58–63, 65, 72, 75, 81, 140; instrument, 18–22,

25–7, 29, 32, 39, 41, 44, 47, 48, 63, 69, 100, 132, 135; listeners, 18, 21, 134; local lines, 18, 22, 44, 65, 67, 93, 98, 100, 101, 107, 108, 113, 115, 118, 123, 131, 135, 137, 138, 142, 143, 145, 150; long-distance lines, 30–3, 35, 36, 48, 142; managers, 24, 29, 92, 95, 103; manners, 91, 108, 133, 134; monopolization, 9, 28, 29, 32, 33, 38, 40, 41, 46, 48; numbers, 25, 97, 115; party lines, 21, 30, 44, 48, 52, 102, 116, 133, 142–4, 149, 152–4, 163; personalization, 28, 102, 108; poles, 23, 24; policies, 14, 17, 31, 32; private lines, 14, 22, 29, 44, 121, 124, 143–5, 149, 152, 168; promotion, 35, 36; public service, 14, 28, 32, 33, 39, 40, 43, 44, 49, 52, 75, 108, 118, 121, 126, 132, 146, 161; rates, 29, 31, 32, 38–49, 110, 116, 118, 121, 124, 144, 147, 152; reception, 18, 20, 21; residential, 31, 37–9, 115, 116, 121, 126, 136, 139, 140, 142, 143, 150; rural, 33, 39, 45–8, 142, 152–4, 160; service, 9, 31, 32, 37, 41–9, 64, 65, 67, 73, 75, 97, 101–6, 118, 121–6, 139, 146, 147, 152, 155, 167; signalling system, 18, 19, 20; sociability, 100, 149, 154, 163, 170–2; social distribution, 5, 8–12, 14, 17, 26, 28, 38, 43, 49, 110–18, 121, 122, 141, 149, 150, 167, 172; social impact of, 3, 10, 11, 15, 17, 18, 25, 96,

155–66; stations, 25, 28, 31, 48, 136, 140, 142, 143, 153; structure, 4, 61, 66, 67, 140, 154; switchboard, 25, 26, 50, 51, 55, 66, 67, 70, 71, 74, 76, 93, 103, 132, 148; system, 4, 6, 7, 9–19, 22, 23, 26–8, 30, 32, 35, 39, 41, 43, 48, 49, 50, 52, 58, 60, 65, 67, 72, 73, 98, 100, 104, 109, 110, 112, 115, 116, 121, 122, 126, 130, 131, 133, 135, 140–3, 145, 146, 148, 154, 158, 162, 163, 167, 168, 170, 172; technological development, 14, 17, 19, 26; transmission, 15, 18–25, 31, 32, 36–8, 41, 45; urban, 33, 45, 47, 142, 143; voice, 66–9, 91–7, 132, 161, 162

Telephone development, conditions of: cultural, 10, 91, 172; historical, 4, 6; ideological, 140, 172; legal, 28, 30, 34; political-economic, 4, 6, 10–12, 15, 26, 28, 38, 39, 41–3, 45–9, 108, 111, 121, 139, 140, 167, 172; social, 6–8, 11, 40, 49, 50, 123

Telephone uses, 3–6, 8–11, 26, 27, 49, 68, 102, 104, 110, 122–5, 130, 136–9, 141, 155, 156, 162, 172; early, 18, 19, 136–9; prescribed, 5, 12, 26, 92, 100, 130–6, 139, 147–50, 154, 159, 170; types of, 8, 100–3, 141; unexpected, 10, 12, 133–5, 142, 152–4; users, 3, 8–13, 17, 18, 21, 25–7, 39, 40, 56, 63, 68, 81, 94–6, 98, 100–3, 108, 123, 126, 130–2, 142, 155, 162

Victorian Act, 107

Walkerdine, V., 5
Watson, J.C., 19
Wharton, E., 160
Williams, R., 15–17, 26, 38, 49
Women: activities, 3, 5, 10, 11, 13, 59, 127–9, 140, 141, 143, 146, 149, 150, 154, 156–8, 170; characteristics, 51–3, 55, 59, 63, 80; as class, 8, 127; contribution to telephone development, 3, 4, 10, 54, 127, 128, 139, 140, 150, 154, 155, 165, 169–72; householders, 51, 127, 128, 140, 146–9, 158, 163, 170; impact of the telephone on, 163–5; lack of power, 5, 8, 135; prescribed telephone uses, 140, 143, 146–8, 150, 152, 154, 170; resistance to prescribed uses, 5, 10, 12, 146, 149, 167, 169–71; role, 3, 11, 12; sociability, 128, 129, 133, 134; social practices, 4, 5, 10, 12, 14, 15, 154, 159, 160; subordination, 4, 51, 54, 59, 60, 69; as telephone users: 4, 10–12, 37, 49, 133–5, 148, 150, 151; unexpected telephone uses, 4, 53, 140, 150–4, 159, 161, 170, 171; uses of the telephone by, 12, 38, 96, 101, 110, 116, 133–5, 139, 143, 146, 148–64, 170; wage workers, 4, 51–4, 128; work, 53, 56